Study Skills
for Criminology

Study Skills
for Criminology

John Harrison, Mark Simpson,
Olwen Harrison
and Emma Martin

SAGE Publications

London ● Thousand Oaks ● New Delhi

First published 2005

SAGE Publications Ltd
1 Oliver's Yard
55 City Road
London EC1Y 1SP

SAGE Publications Inc.
2455 Teller Road
Thousand Oaks, California 91320

SAGE Publications India Pvt Ltd
B-42, Panchsheel Enclave
Post Box 4109
New Delhi 110 017

British Library Cataloguing in Publication data

A catalogue record for this book is available
from the British Library

ISBN 1 4129 0322 X
    1 4129 0323 8

Library of Congress control number available

Typeset by C&M Digitals (P) Ltd., Chennai, India
Printed on paper from sustainable resources
Printed in Great Britain by The Cromwell Press Ltd, Trowbridge, Wiltshire

Contents

Preface

This book has been written to take account of some of the significant changes that are taking place in Higher Education: widening participation, a greater emphasis on vocational programmes, a focus on employability skills and the introduction of Foundation Degrees have all had an impact on the student experience at university, and on some of the learning and teaching strategies employed by lecturers and tutors.

Current government policy is aimed at increasing the number of young people who undertake programmes of study in Higher Education. While there may well be debates about whether this is a good thing or not, the reality is that many of you who are studying at university now, or are looking to begin studying in the future, are from a much wider variety of backgrounds than was the case twenty or thirty years ago. While most of you will have some experience of studying before beginning a degree programme, the practices that are common within Higher Education Institutions can often be very unfamiliar to you. This book has been written with these changes in mind and is designed to provide a guide to studying at university with a subject specific focus on Criminology. The intention is to provide you with the basis for developing good study habits and to enable you to get the most out of your programme of study while making the most of your experience of university life.

It is worth asking, 'Why do we need to produce a study skills guide?' given that access to Higher Education is usually based on some level of previous study and academic achievement. In many universities, lecturers and tutors are identifying areas of the curriculum that students successfully understand and are able to discuss appropriately but they do not seem able to present this as successfully in written assessments. A guide designed to develop skills for this level of study will therefore be useful. Equally, there are an increasing number of you who are becoming the first members of your family to become students in Higher Education. This means that there are many of you who may be unfamiliar with the nature of university life, and this can often make adjusting to the greater emphasis on 'self-managed' learning that is expected in Higher Education difficult. Again, a guide addressing some of these issues can help to smooth the transition from school/college or work into Higher Education. In addition, there are many of you who are mature students returning to full-time education who have limited or no recent educational experience and others who choose to study on a part-time basis. The time-management skills and guidance on assessment planning in this study guide will, therefore, be of value to you during your degree.

Another factor that is influencing the way you are taught and assessed at university is the increased emphasis on employability skills: there is no longer an assumption that by gaining a degree you will have gained the skills that employers require. The University Vocational Awards Council (UVAC) and Skills for Justice, which is the sector skills council for the Justice sector, are both focusing their attention on raising the importance of vocational skills and graduate employability. The impact of this on degree programmes also has an influence on your experience as a student and this book seeks to achieve three outcomes to enable you to have a positive experience of university:

1   to raise your awareness of the range of academic study skills necessary to gain a degree in Criminology

2   to provide guidance about the ways in which you may be assessed in these skills

3   to offer advice on how you can demonstrate your skills in a way that will meet the needs of employers.

Introduction

The aim of this book is to provide a guide to the general study skills needed in Higher Education and then apply these to the subject area of Criminology. It is not the intention here to provide an in-depth discussion of criminological debates, theoretical concerns or **criminal justice** policy, but simply to provide some examples and direct you to other literature that develops the debates more fully. The main skills identified here are to do with:

- **researching the topic and data collection**

- **presenting the data for assessment purposes**

- **using resources effectively**

- **learning to reference correctly**

- **examination techniques**

- **developing transferable skills**

- **the transition from education to work**

- **ensuring that you get the most out of your degree and the experience of University.**

The book seeks to demystify the world of Higher Education and so provide you with the skills necessary to make the most of your time at university. At this point it is worth noting some key terms that are used to describe the way that degrees are organised; these terms are used interchangeably in different universities but have the same meaning and we need to clarify, at the outset, the terms we use within this book.

Programmes or courses

A 'programme' is the degree that you are studying, BA/BSc Criminology, for example; however, some universities use the term 'course'. In this book we use the term 'programme' to refer to the degree you are studying. A 'programme of study' is made up of a number of 'modules' or 'units'.

Modules or units

Modules are the individual units that make up a programme. They are the subject areas that you will study. As will be discussed in Chapter 2, you need to accumulate passes in a number of these to gain your degree. In this book we use the term 'module'. You will find some examples of the type of topics that might be studied in Chapter 6 (see also Appendix 3).

Using this book

This book is predominantly aimed at undergraduates and will perhaps be most useful for those of you who are in the early stages of your studies. It is an especially useful tool for those of you who are part-time, and for mature students who have to balance a number of competing commitments, especially when returning to study after a long time outside the academic environment. However, there are sections of the guide that will be of use throughout your undergraduate career; for example, the advice on examinations, essays and other forms of assessment that you are likely to undertake throughout your programme can be returned to whenever assessments are due. The way to get the most from this book is to refer to specific chapters at appropriate stages of your academic career, for example, when beginning to write an essay, or shortly before you take examinations or are required to make a presentation. The book has been written in an academic but accessible style so that you can develop your skills over the course of your degree programme. The content of the guide is based on the collective expertise of the authors who have a combined total of over thirty years' experience in teaching study skills in Higher Education. However, it is not the intention that you follow the guidelines rigidly, but that you take from the book the information and advice that suits your own style of learning; you will all develop your own style while at university and to some extent this will be dictated by commitments you have in addition to studying. The evidence suggests that those students who are the least organised are the ones who fail to reach their full potential. This book then provides a framework that will enable you to reflect on your own learning style as well as your lifestyle, and provides advice that will help you to achieve a good level of academic success.

The book has been designed to provide examples of important facts and information, to present activities to help in skill development and to give useful key tips on studying. Where necessary key terms are explained to help you to understand some key subject specific terms, academic language and, in some instances, the jargon used in universities and in the discipline of Criminology.

Content of the guide

The main content of this book is divided into three parts and these can be used independently or in conjunction with each other. The book begins with more general points relating to study at university and progresses to the development of study skills before engaging in some criminological discussions and subject specific advice. Throughout the book each chapter focuses on various aspects of your degree programme, but, of course, these all relate to each other and where appropriate we make the necessary connections. Some issues are mentioned several times in different chapters and these will be some of the most important things to remember as they are the areas where marks are often lost or students do not prepare sufficiently. Throughout the text we present key tips, which you should find useful and in the Appendices there are examples of writing styles, module guides, assessment feedback sheets, as well as a range of useful websites and contacts.

In Part 1, the four chapters look at studying in Higher Education. Chapter 2 is a guide to the university experience and covers topics such as what it is like to study at university, how this differs from other learning experiences and what is expected of you as a student. It also explains the academic year and how your formal learning is structured. The exercises in this chapter are designed to help you think about the degree you have chosen to study and to assist in making decisions about what you hope to get from the degree. Chapter 2 also includes techniques to help you to manage your time more effectively. This is something that is becoming important for many of you as it becomes increasingly necessary to juggle study, work and family commitments during your time at university.

This chapter also introduces you to university life, especially from the point of view of the academics you will study with and what their roles are.

Chapter 3 looks at assessment, explaining why you are assessed and how that assessment is carried out. There is a wide range of assessments that are used in Higher Education and examples of these are provided. A key feature of this chapter is the focus on what is expected of you and a general guide as to how your assessment might be graded.

Chapters 4 and 5 discuss some of the specific forms of assessment that will be used to monitor your progress and achievements during your degree programme, and will provide advice and guidance on how to prepare for examinations, presentations and other forms of assessment that are commonly used in Higher Education. The activities that are included have been designed for use in conjunction with the actual assessments you could be undertaking on your programme and they will provide the basis for developing your own study skills as you progress through your degree. You may also undertake some of these activities before you receive your actual assessment as they can help you to understand which skills you need to develop to be a successful student before completing work that will be graded.

In Part 2, we begin to look more specifically at the study of Criminology introducing you, albeit briefly, in Chapter 6 to some discussion of what Criminology is. The theoretical and philosophical debates surrounding this area are the subject matter of many texts and, as such, are too substantial and far ranging to discuss in the context of this book. We do direct you to some of these debates and, throughout your degree programme, as you develop your own ideas about these, you will more than likely find yourself supporting some theoretical positions more than others based on your learning and personal views about crime and disorder. Chapter 7 directs you to a range of resources and explains some of the ways in which you can gather information to support your studies. While formal lectures provide you with a signpost to specific topics, to fully understand the issues that are raised you will need to do much more independent reading and research, as simply attending a series of lectures will only give you a minimal understanding of the programme and subject area. In addition to suggesting ways to organise your notes and study, this chapter also discusses the concept of reading for a purpose, and this is also a feature of Chapter 8. The activities included here will help you with this. The chapter also draws your attention to the variety of resources that are available to you. Advice is given on using the Internet and other electronic data bases as well as ensuring you know how and when to reference (see Box 1a) your work correctly. In this chapter we provide you with a range of useful sources of information; these will prove invaluable as your degree progresses.

Box 1a Referencing

This is discussed in depth several times throughout this book (see, for example, section 8.6) and will be mentioned constantly throughout your academic career. It refers to the way in which you cite your sources when writing in Higher Education. In other words, it is the process by which you tell the reader where you got your information from and acknowledges that you are using the work of another person.

Given that you are required to study independently for much of the time you are at university, finding relevant information is an important aspect of the degree. Chapter 7 focuses on this and here we discuss the wide range of resources and sources of information available to you including, textbooks, books based on research (monographs) and electronic sources of information such as the Internet. This chapter also highlights a number of useful websites related to crime and criminal justice, and gives advice on how to research for your assignments. This is not an exclusive list as the amount of material available in this way is too extensive for a book of this nature; it does, however, provide a good starting point for your research. We also try to demonstrate the reasons why you need to support your discussions with evidence and explain how to do this. Chapter 8 looks in some depth at essay writing as, increasingly, in Higher Education essays form a significant element of the assessment process, and, in some instances, this may be the only assessment for a module. We use this chapter to develop your understanding of what an essay is and how you should prepare and research when writing for academic purposes. There is also advice about the structure of essays and exercises on time-management strategies that will help you complete your research and produce your essays by the deadlines set for submission. This chapter also includes advice on preparing, making and writing up presentations; these are forms of assessment that are used increasingly and provide you with the transferable skills relevant to employment opportunities.

In Chapter 9 we provide you with an introduction to criminological theory. Following this we explore the links between these theoretical perspectives and a selection of criminological topics. (While most are fairly straightforward we do attempt to develop a more complex understanding of this theme to provide a greater challenge.) Related to this, the chapter introduces you to the key skills of comparing and analysing different theoretical perspectives. This is an important issue as your degree will be delivered in modules or units (as described in Chapter 2) and if you see each module as freestanding you miss a real opportunity to develop your expertise and breadth of understanding in this discipline.

Finally, in Part 3, we reflect on the content of this book and develop some understanding of how you might use the skills and knowledge that you have gained on your degree after you have graduated. Chapter 10 reflects on our discussion and highlights those areas that will help you to show your own skills. Employers often see the skills you acquire while studying at university as 'employability skills' and there is evidence (UVAC, Skills for Justice) that graduate skills are highly sought after by an increasing number of employers. As a consequence we link the book to the world of work by discussing how you can make the degree work for you. This important aspect of studying in Higher Education is the focus of Chapter 11, which considers the ways in which you can use your degree to document the evidence for the skills you have gained, and how they are relevant for future employment and career aspirations. We also introduce you to a selection of occupations that you may wish to consider once you have completed your degree programme in Criminology. Obviously, this cannot cover every possible vacancy and career opportunity and is only intended to give you a broad idea of the range of options that may be available to you after graduation. One specific feature of this chapter refers to the introduction of Personal Development Planning (PDP) into Higher Education. This is something that will be introduced universally at the beginning of the academic year 2005/06, although some universities have already developed and introduced this. While the way in which this will be facilitated varies between institutions the content in Chapter 11 is intended to demonstrate the value of engaging with the process and also to highlight the reasons why it is increasingly considered useful by potential employers. You are invited to look at the skills you already possess as an individual when you begin your degree and to relate these to the occupations you might envisage working in after graduation. One of the important features of this self-analysis of your skills is the fact that as you progress through your programme of study, your abilities in certain key areas will develop and the maintenance of a PDP is one way in which you can record your own individual development, both academically and personally.

Here, too, you will find advice about the compilation of a curriculum vitae (CV), which, in addition to presenting factual information about your education, work experience and relevant personal information, can be a summary of the key skills you have acquired and relevant aspects drawn from your PDP that you can relate to specific job applications. The activities in this chapter can provide the basis for both of the processes mentioned here and will be made use of in a variety of ways for different situations, depending on the purpose for which you are using the information and how it is being presented.

Part One — Studying in Higher Education

Part 1 looks to set the scene for you and to discuss some key issues related to studying in Higher Education. This is important because the majority of you beginning a degree at university for the first time will do so with an increasingly varied range of previous experiences. While many of you will no doubt have undertaken traditional 'A' levels, an increasing number of you will have gained entry onto your programme with a wide variety of alternative entry qualifications, such as Access Courses, National Diplomas and vocational qualifications.

To help you to make the transition from school, college, part-time study or work, this section of the book addresses a number of the areas that you may not be familiar with, and introduces you to the variety of assessments you will encounter during your time at university.

The four chapters in this section will look at:

- **The Academic Environment**

- **Assessment in Criminology**

- **Presentation Skills**

- **Examination Assessment.**

2 The Academic Environment

CHAPTER OVERVIEW

By the end of this chapter you should be familiar with:

➤ the structure of the academic year
➤ the modular structure
➤ teaching in Higher Education
➤ the nature of the academic environment.

2.1 The academic year and modular structure

Studying at university will be a new experience for the majority of you who are beginning a Criminology degree; however, studying in general should be familiar to most. While the academic year usually runs from September to June, as it does at school and college, some elements of the academic year may be structured in a different way to what you are familiar with.

The academic year in Higher Education is normally divided into either three terms or two semesters. Some universities have retained the traditional three-term year, which most of you are probably used to from school or college, while others have adopted a semester structure. Those institutions using terms usually follow a similar pattern to the school year in England and Wales, with each academic period being related to Christmas and Easter. It is worth noting that in many universities there are no mid-term breaks although some do incorporate reading weeks; these are when lectures and seminars do not take place to allow you the opportunity to reflect and read the recommended texts that will develop your understanding of topics you are studying. Semesters, on the other hand, run from September to January and from February to May cutting across these religious festivals, although there continues to be breaks for both. The majority of universities will begin the academic year during September or early October and this is when you will begin your programme of study. In some instances there are other start dates, for example in February at the start of Semester 2 but this is not often the case. Although the academic year normally ends in June, at many universities you will not have any formal contact, such as lectures and seminars, after the end of May as this is when most of the programme administration and examination boards take place.

TERMS

Most of you will be familiar with terms as the traditional school year in England and Wales is divided in the same way. Terms run as follows:

Term 1	September to Christmas
Term 2	Christmas to Easter
Term 3	Easter to May/June

In universities when the academic year is divided into terms you will either study modules or units over the course of the full academic year, or the modules will run for the duration of the term.

SEMESTERS

Semesters have been common in many universities since the middle of the 1990s. They are linked broadly to the traditional academic year but do not end at the usual Christmas and Easter breaks, however most Higher Education institutions take a break at these times. Semesters run as follows:

Semester 1	September to January (15 weeks excluding Christmas break)
Semester 2	February to May (15 weeks excluding Easter break)

For those universities that divide the academic year in this way it is more likely that you will study modules in two blocks of 15 weeks, including assessment, that is time set aside for the submission of course work and taking examinations. For example you may study three modules in Semester 1 and three in Semester 2, and again it is important to remember that these are related to each other and not individual stages of your degree that are simply to be overcome.

HOW IS THE TEACHING STRUCTURED?

Whichever structure is in operation within the university you attend you will be taught a number of modules (see Box 2a) or units covering specific aspects of the programme throughout the academic year. Throughout this book we will use the term 'module' but this refers to both modules and units. These terms are used interchangeably within Higher Education but many of you will be familiar with modular learning from school or college.

Box 2a What are modules?

Modules are courses relating to a specific topic or subject area; as mentioned earlier, they usually last for either one semester (15 weeks) or one term over the full academic year.

Each module will have a credit value and the accumulation of these credits will lead to the award of a degree.

As an example, consider a Level 1 module in Criminology, let us call it Justice and Society, with a value of 20 credits. This may run during Semester 1, September to January. You could perhaps have 24 × 1 hour lectures supported by 12 × 1 hour seminars with an assessed essay to be submitted at the end of the semester. You need to gain a pass mark for the essay to be awarded the appropriate credits.

In the example in Box 2a you would need 120 credits to pass the year and so you would study three modules with a 20 credit weighting each semester; all will have a similar format. At most universities you are required to gain 120 credits at each of the three levels you study.

Levels relate broadly to academic years so first year is Level 1 and so on when studying on a full-time basis. For those of you who choose to study on a part-time basis it is possible that you will be studying different levels at the same time although when you begin your degree you will take Level 1 modules as the requirement for your future study. Modular credits are usually 20 credits although some institutions vary this to 10 or 12. If your institution uses 20-credit modules you will study six modules in each academic year, 10 credits equals twelve modules and 12 credits would equate to ten modules. Modules that have to cover a lot of material may be double modules carrying a double weighting of credits. So for example the final year dissertation or a Level 1 module covering major themes on Criminal Justice that will be an essential part of your degree may be worth 40 credits. Each module is assessed independently using a variety of assignments, which could include essays, projects, presentations, case studies and exams; these will be discussed in greater detail throughout this book. The modules you are required to study will usually be two types: core modules and option modules.

CORE MODULES

These are compulsory modules that you must study and pass in order to be eligible for a named Criminology degree to be awarded to you; these will be seen as the basis on which your understanding of the discipline is developed. Typically core modules will be in the area of criminological and social theory, the criminal justice system, and systems of crime or social control, punishment and research methods. Depending on the overall structure of your programme some institutions may see Psychology, Politics, Social Policy or Policing as key and there may be core modules in these areas. In your final year you will usually be required to produce a dissertation or research project in order to be eligible for the award of an honours degree and this will also be seen as core.

OPTION MODULES

These are modules that you can choose depending on your interests. This will not be a totally free choice as the options offered usually relate to the research interests or specialist knowledge of the lecturing staff that teach you. You will, however, be able to have some choice that allows you to develop an interest in a specific aspect of criminological study and to this end your dissertation or project in your final year can build on this. It is, therefore, quite important to give some thought to the options you are choosing to study. It is not a good idea to choose something because the assessment seems easy, for example, not having examinations. Most people would prefer not to do examinations but if you have a real interest in a subject you will perform much better in whatever assessment you do rather than in a topic that you don't find stimulating.

Activity 2a may be useful when deciding what options or specialist areas to study in your degree. Appendix 4 also gives a sample module guide.

Activity 2a

Make a list of the different areas you are familiar with in the degree programme you are studying, for example, policing, domestic violence, youth and crime, drugs and crime, gender and criminal justice, race and ethnicity, crime prevention and the criminal justice process.

Once you have written your list score the topic areas in relation to your enthusiasm for them: score one point for those that you enjoy and feel confident in, two for those you enjoy but find difficult, and three for those you enjoy the least.

Where a module relates closely to areas in which you have scored one you will be making a good choice as this is a module you are likely to enjoy and succeed at. Those scored two may be enjoyable but you will have to make more effort to succeed; however, this would still be a sensible choice. For those scored three you are not going to enjoy the module and may struggle to produce work of a good standard.

It should be noted that this is a generalised way of choosing options and it may be that you have a career plan that will require you to study modules that you find difficult, or less interesting. It is also a fact that some modules may seem less interesting before you begin to study them but the content, when you get into it, is stimulating, so it is important when making your decisions that you speak to staff who can advise you about the module. The golden rule here is not to leave option choices to the last minute; you need to think about them well in advance.

2.2   How will the modules be taught?

At university, modules are usually taught by a combination of lectures, seminars, tutorials and workshops that require personal, individual or self-directed study. Some of these terms are used interchangeably and you will become familiar with the ways in which these terms are used within your own institution very quickly.

LECTURES

Lectures usually take place in a formal teaching setting; normally a lecturer will present some material on a specific topic in a lecture theatre to all of the students studying the module and, unless the lecturer specifically invites comment, the normal etiquette here is to remain silent and take notes as the lecturer has to present a certain amount of information in a short space of time. The way in which lectures are presented varies and will be in a style that suits the lecturer. For this reason it is not likely that you can expect all lectures to be in exactly the same format, or for all lecturers to provide you with the same supporting material. Some lecturers may use an overhead projector to present slides outlining key points, while others may prefer to use PowerPoint if this is available; indeed, some lecturers may prefer to use neither. It is important that you acquire skills that enable you to listen to the lecture while making note of the important points. While it may be that you are provided with an outline before the lecture or you may be given a handout containing relevant material at the end of the lecture, it is important for you to take notes to remind you of the key points. Usually lecturers will provide you with reading related to the topic either in the lecture or in the module guide, which you will have received at the beginning of the module (an example of a module guide is contained in Appendix 4). It is important to follow up the lecture by reading at least the key texts; this helps to develop your understanding of the subject and provides the basis for completing assignments and taking examinations. Reading for academic purposes and undertaking assessments are some of the key points that are discussed at length throughout this book; these are essential tools that will enable you to maximise the grades you get for your assessed work.

SEMINARS

These are less formal learning areas and usually involve smaller groups. They are more like the classes you may have been familiar with in school and college and allow you the opportunity to discuss the lecture material in some depth. This is when your own reading is important because you will be able to contribute much more if you have prepared well. A member of academic staff will facilitate the group but there is an emphasis on students making a significant contribution. The reading you do following your lecture will enable you to make a positive contribution to the debates; it is important that you do not feel diffident about challenging the views of others, although it is essential that you support your views with evidence from the reading you have done. (The importance of evidence in academic writing is discussed in more depth in Chapter 6.) This is not the place for populist debates as expressed in the media or for engaging in 'casual' conversations about the nature or extent of crime; you need to be developing academic arguments about the discipline. Within seminars you may be required to present discussion papers focusing on a specific question or issue; this is intended to be the basis for discussion within the group and often helps in the preparation of essays. It is also becoming more usual for groups of students to make presentations within seminars that are assessed and this helps to develop your skills in gathering and presenting information in a structured way. Presentations are discussed in more depth in Chapter 4 but it is worth emphasising at this point that the skills involved are recognised as key transferable skills by employers. It should be noted that in some universities seminars are referred to as tutorials although the aims and general structure are the same.

KEY TIP

Always read the key texts before attending seminars; you will develop your understanding of the topic much more effectively if you are prepared.

TUTORIALS

These sometimes take the format described as seminars, as discussed above; however, it is likely that you will also be required to attend individual or small group tutorials for academic support, personal development or pastoral support. This is something that will become increasingly important as universities, from 2005, are required to provide the opportunity for you to develop a Personal Development Portfolio (see Chapter 11 for a full discussion of this). Most universities will allocate you a personal tutor who will support you academically and will help you with any non-academic difficulties that you may encounter during your programme. It is usually expected that you maintain contact with your personal tutor even when you feel that things are going well! The personal tutorial system is different in each institution with some allocating a personal tutor for the duration of your programme of study, that is all three years, while others allocate tutors to each year of your programme. There will also be other arrangements depending on the way a programme is structured. The tutorial system will be described in your student handbook, which you should receive at the beginning of your programme.

WORKSHOPS

These are increasingly used within Higher Education and allow you the opportunity to work in groups dealing with some of the practical issues related to your degree. These differ from seminars in that there may be several groups working on different projects or case studies and there is unlikely to be a summary of your work at the end of the session. This is because you may be asked to use a workshop to develop a presentation and this could in some cases involve working in a group while seeking advice from your module tutor or workshop leader. Workshops are also used as the forum for discussing and developing research projects; here, you may have the opportunity to discuss the research methods you want to use and the way in which you can present your proposal and findings. You will also find this is where you may be asked to work on case studies which allow you the opportunity to apply criminological theory to criminal justice and **crime control** practice as discussed further in Chapter 9.

PERSONAL TUTORIALS

These are often delivered on a one-to-one basis although some sessions may be group based. This is when you will meet the tutor assigned to support your academic and personal development throughout your degree programme. You may not have many formally scheduled meetings for personal tutorials, rather, you will be able to request an appointment for support when you need it. Most universities ask you to see your personal tutor at least once each term or semester to monitor your progress.

KEY TIP

Make good use of your personal tutor. They can help you make the most of your time at university and often are required to provide references when you leave. The better they know you the more useful their reference will be!

It is important that you realise that your personal tutor will not always have answers to your problems, but they will be able to direct you to those people who have experience in dealing with specific issues, such as accommodation, finance or health. These will be arranged so that you can have a one-to-one appointment to discuss your individual circumstances.

2.3 How much time will you be expected to spend studying?

The amount of time you are required to study is one of the most difficult things to get used to at university. The actual time you spend as a full-time student in lectures, seminars and tutorials may not be much more than ten or twelve hours a week and, indeed, some universities may well require even less contact time. The amount of actual 'learning' hours that you are expected to devote to each module, however, will be much higher. This is where you begin to make use of your time-management skills and when you need to develop the ability to be a self-motivated student.

KEY TIP

Typically for each hour of contact time with academic staff you will be expected to add at least two hours of self-directed study, reading, writing notes, preparing essays and preparing for examinations.

This is becoming one of the most difficult areas for students as they try to juggle a number of competing demands at the same time. In many universities it is increasingly being reported that students who are considered to be full-time students are in fact balancing work as well as family roles with their studies. This variety of competing commitments requires the development of disciplined and high quality time-management skills. This means that you really do need to keep a diary, (an academic diary that covers the academic year rather than one covering the calendar year is ideal); this ensures that you are able to keep a close check on your commitments. It is necessary that you begin to diarise your social, working and academic life to ensure that all of your targets and deadlines are met; if you do not do this you will find yourself falling behind with your work. It is important to remember that the time spent outside lectures and seminars or tutorials can be as stimulating academically as the time spent in these more formal aspects of study. It is, therefore, a good idea to build into your schedule the opportunity to meet socially with your student colleagues. It is often in these situations that some of the best ideas are developed and the debate is at its most exciting. The opportunity to develop your understanding of the topics is immense and you can test out some of the perspectives you are learning about. A cup of coffee, a glass of wine or a bottle of beer can be the perfect companion to a lively debate on the criminal justice system or on theoretical perspectives on crime and criminality.

Activity 2b

Using the grid below as an example draw up a schedule of your weekly commitments, including those for academic study, lectures, seminars and tutorials, work and family. Also include your social or leisure activities. This effectively becomes your personal timetable.

	Monday	Tuesday	Wednesday	Thursday	Friday
9–10					
10–11					
11–12					
12–1					
1–2					
2–3					
3–4					
4–5					
5–6					
6–7					
7–8					

You may need to amend this grid slightly to fit with your own circumstances. It is worth noting that this is especially important for part-time students and those students returning to study after a number of years.

Time management is one of the most important skills you will develop at university as you will have periods of time during the academic year when there appears to be little or no pressure to complete work. You are unlikely to be asked to produce 'homework' in the way that you might have been at school or college. It is also important to note that the academic staff may be less freely available than your teaching staff at school or college might have been. This is because of the wide range of additional roles that they often have such as administration, research, and contractual work outside the university.

2.4   Who are your lecturers?

While most of your lecturers may not necessarily be seen as 'the big names' in the discipline, it is likely that many of them will have published research papers, research reports or presented conference papers. Most academics are actively involved in research and publication that contributes to the understanding of crime and criminal justice. In some cases the textbooks that are recommended may have been written by some of the academics you come into contact with. It is important to remember that this activity is not in addition to the work that lecturers do in the university but is seen as an integral part of their role. This mixture of research and teaching is particularly important in Higher Education as the issues that are discussed in lectures and seminars can be informed by current research and are therefore based on the most up-to-date information available.

It should also be noted that as part of the quality process in Higher Education all institutions employ lecturers from other universities as external examiners. This often involves people being out of university for a few days on occasions, which means that they are less accessible to you at these times. In addition, lecturing staff attend and contribute to conferences and engage in a variety of scholarly activities that mean their availability is not always open access on demand. It is worth noting that academics will respond positively to your enthusiasm for the discipline and will be keen to discuss the topic with you if you are informed and well read. It is good practice to make appointments and keep them, to communicate effectively and to ensure that, at an early stage, you inform the university and appropriate tutors or programme leaders of any problems that you are having. This allows staff to support you in your progress and ensure you make the most of your time at university.

2.5   How should you study?

Adapting your learning style to meet the greater flexibility and freedom of Higher Education is one of the more difficult adjustments that new students have to make. This is particularly the case when you have been used to a very structured style of learning. The construction of a personal timetable (Activity 2b) is important in providing yourself with the discipline necessary to study effectively and successfully. It will also allow you to develop your assessment in an ordered way as we discuss in Chapter 3.

Reading for a degree is a central feature of university study and you must support all of the taught content from lectures and seminars with your own reading based on the recommended material presented by lecturers and tutors. Although your lecturers will direct you to key reading, over time you should expand this as your own interest and understanding of the subject material increases. As a minimum you need to read the recommended literature, but the more you read, the greater depth and breadth you can include in your assessed work. There is a skill to reading for academic purposes and this involves note-taking while you read. If you simply read an academic text in the same way as you would read a novel you may fail to grasp many of the key issues included in the discussion. Hopefully your reading will raise questions in your mind and you need to write these down so that you can expand your thinking as your understanding of the topic develops. This is a key part of the process by which you begin to practise analytical skills, which improves the quality of your assignments. This enables you to engage in the debates, while demonstrating your understanding of various perspectives and discussions in depth.

2.6   What will your time at university be like?

You should find your time at university enjoyable and if you study properly you will make this an exciting and pleasurable time of your life. Those who structure their work can build in the various additional activities that help you to have fun while working hard. It is when you come under pressure that you feel you cannot control that the greatest problems occur. Students often claim to work 'better under pressure'; however, this does not mean leaving everything until the last minute!

You should build into your schedule opportunities to mix with other students, join student groups and societies and engage in activities outside your degree programme. Some universities now have links with local groups that allow you the opportunity to do voluntary work; this can be a benefit when you begin to look for work. Other examples include sports clubs and groups who share an interest in drama, music, debating and other specialist activities.

2.7   Coping with anxiety

Throughout the course of your degree programme there will inevitably be times when you feel anxious about your progress, your workload or about situations outside university. This will frequently be related to assessment and in some cases to personal situations. While it is not the intention that we tell you how to resolve anxiety, or attempt to provide some form of counselling, using some of the advice given in this book can help you to deal with and prevent some of the situations that contribute to anxiety. This will help you to have a positive learning experience during your time in Higher Education. For most people there are times when they feel anxious because of circumstances that may or may not be of their own making. Even when the source of the anxiety is beyond your control there are steps that you can take to minimise the effect of these situations. There are many sources of information and support related to stress. Most universities will provide support and many have websites to give useful advice; one such site is sponsored by the University of Leeds and is called ahead4health. This site can be found at: http://www.leeds.ac.uk/ahead4health/help.htmsite and Box 2b lists some of the topics covered.

Box 2b Dealing with anxiety

Some of the subjects covered in the ahead4health website include the following topics that can affect students during their time at university:

- alcohol and drug use
- anger management
- anxiety
- appearance — worry about ...
- bereavement and loss
- depression
- eating problems
- exam anxiety
- feeling homesick
- insomnia
- putting things off (procrastination)
- rape and sexual assault
- relationships and sexual problems
- sexuality
- helping a friend in a crisis.

It would be nice to think that students when they come to university could spend three years studying and making friendships and relationships without any worries; however, it is possible that this will not happen. Although some of you may suffer only minimal disruption it is likely that there will be unforeseen circumstances that will disrupt even the most well-structured routines. Your personal tutor is one of the first people you need to contact when any kind of problem occurs, whether you anticipate that this might happen or when it is unexpected.

2.8 Summary

This chapter has hopefully demystified the world of Higher Education and the key themes covered have been:

- the structure of the academic year

- who the lecturers that teach you are

- demands on your time

- coping with anxiety.

Finally, considering some of the 'dos and don'ts' of university life listed in Box 2c may help you to reduce some of the potential anxieties that can occur.

Box 2c University life — dos and don'ts

- **Don't** think you are the only person who is confused, or misunderstands. Especially in the first few weeks at University, remember that everyone else will be experiencing the same uncertainty and nervousness.
- **Do** be bold and speak to people, express your concerns and when you are confused **ask for help!**
- **Do** organise your time. As we have discussed earlier in this chapter university is very different from school or college. In relation to work you will be left to your own devices; you need to motivate yourself.
- **Don't** hand work in late — late submissions receive zero grades.
- **Do** speak to someone if you have a genuine problem about meeting deadlines for submission of assignments; most universities will have a formal system to deal with this.
- **Don't** allow yourself to become isolated. The wide range of interests that are catered for means that there will be groups and societies for people with the same interests, for example, sports, leisure interests and special interest groups.
- **Do** approach your student union to set up a group if your interests are not represented. There are likely to be other students who will be happy to join.

The major don't

- **Don't** copy other people's work and claim it as your own. This is what is referred to as plagiarism — it's as if you went into a supermarket and put stuff in your bag without ever intending to pay for it. You can use the work of others to support your work as long as it is correctly referenced (see Chapter 8). Generally speaking, you will be treated progressively more severely depending on the extent of the plagiarism, your programme level, your previous learning environment (how familiar you might be with UK academic conventions) and whether you have been guilty of plagiarism previously. The penalties range from a reduced mark to your work receiving an outright fail. Furthermore, some professional bodies identify this as dishonesty resulting in either a failure to award an accredited degree or exclusion from membership of that body, which is likely to affect your career opportunities.

3 Assessment in Criminology

CHAPTER OVERVIEW

By the end of this chapter you should be familiar with:

➤ different types of assessment used in Higher Education programmes of study
➤ expected learning outcomes
➤ the importance of peer and self-assessment
➤ choosing modules.

3.1   What is assessment?

The Quality Assurance Agency (QAA) checks quality and standards of programmes and subjects within all Higher Education institutions by auditing them. The QAA Code of Practice on the Assessment of Students states that:

> Assessment is a generic term for a set of processes that measure the outcomes of students' learning, in terms of knowledge acquired, understanding developed and skills gained.

and that:

> It provides the basis for decisions on whether a student is ready to proceed, to qualify for an award or to demonstrate competence to practise. It enables students to obtain feedback on their learning and helps them improve their performance. It enables staff to evaluate the effectiveness of their teaching. (www.qaa.ac.uk)

Whether it takes place during a module or at the end of the module, and no matter whether it consists of an essay, presentation, examination or portfolio, assessment is an important issue for students. This chapter discusses types of assessment that you may come across during your university career and explains why this assessment takes place and the learning outcomes involved. Chapters 4, 5 and 8 will attempt to provide some advice for you as you negotiate your way through assessment with notes and tips on essay writing, revision, examination techniques and strategies, presentation skills, report writing and portfolio-building ideas. These practical examples from within the Criminology discipline, as well as the activities that you can complete, will enable you to find the best way of studying Criminology whether you are new to studying or an experienced student.

3.2 Types of assessment — formative and summative

You will be assessed within the time that your programme takes place or at the end of the course. This work may be:

- **formative**

- **summative.**

FORMATIVE ASSESSMENT

This does not contribute to overall assessment and grading is not necessarily given for this work. Where a grade is given it is usually of a formative nature and does not usually count towards the final grade for the assessment on the module. This is an instance when peer review sometimes takes place. The rationale for asking you to complete formative assessment is because it provides you with feedback on your progress and informs your future learning. For example, you may be asked to produce a ten-minute presentation for a seminar class but will not be marked on that piece of work. Instead, the presentation would lead the other students in the seminar group to focus on a specific aspect of Criminology while you will gain the practice of preparing and delivering a presentation. An exercise like this will also allow you to receive feedback from the seminar tutor and possibly, as mentioned above, from your peers. There will be less pressure on you to perform in order to receive a grade but you will be able to develop skills that will help you to achieve a higher grade in assessed work. Formative assessment is useful in identifying areas which require further work and such presentations are especially useful for you during your first year as an undergraduate, although if too many formative pieces of work are given to students this can add to the burden of your workload. For this reason lecturers and tutors will be selective in asking you to carry out this type of assignment.

SUMMATIVE ASSESSMENT

This provides a measure of achievement made in respect of a learner's performance in relation to the intended learning outcomes of the unit, module or programme of study. One example of summative assessment would be an examination at the end of the module. This would check that you have achieved the learning outcomes of that module and would be marked and graded. Summative assessment may also consist of essays, reports and presentations that are graded and that count towards your final mark for the module. This would also mean that you would gain credits in order to progress from Semester 1 to Semester 2 or between years. As discussed in Chapter 2, you are expected to gain a specified number of credits during each academic year to enable you to progress to the next year of study.

3.3   Peer assessment

As previously discussed, presentations may often provide the forum for peer assessment. This is a form of assessment used in Higher Education that may be new to many students. Peer assessment is an evaluation of your work by your fellow students. In this form of assessment you may be required to grade another student's work. This could be an essay plan, a short essay or a presentation and the mark you give may form part of the overall grade for that module. An example of this might be that your peer group will appraise the ways in which you as a presenter have clearly identified and delivered the key points and issues relevant to the topic. They will also identify how well your presentation stimulated discussion and raised questions at the end of the presentation. This method of assessment can be extremely useful as, if you are required to appraise someone else's work, you need to know the subject area in some depth yourself! Evaluating other people's presentations in terms of, for example, style, structure and content also enables you to be able to pinpoint the good and not so good parts of your own work. An example of a Peer Assessment Mark Sheet is given on page 29 and highlights the areas that are assessed.

Peer Assessment Mark Sheet

Give each category a mark from 1 to 5 with 1 being the lowest mark and 5 the highest:

	[1]	[2]	[3]	[4]	[5]
Clear plan					
Key issues identified					
Logical structure					
Arguments made sense					
Resources used: OHPs Powerpoint, Handouts, Flipchart, Video					
Teamwork					
Timing					
Promoted class discussion					
Entertaining					

Other comments:

Names of students presenting:

Topic covered:

Presentation date and time:

Peer assessment is not only an extremely valuable exercise within Higher Education but is also an important skill in relation to the workplace. In a general sense, when we attend meetings, make presentations or take part in training sessions, for example, there is a need to produce relevant information in a coherent manner so that people listening to you can understand your intended key points. Most jobs and professions within the world of work are service based and being able to present reports that are well structured and clearly written is a skill highly valued by employers, colleagues and clients. Taking part in peer assessment during your university degree can give you the confidence to be able to respond positively to points raised by your peer group rather than feeling exposed, resentful or threatened by a perception that someone has challenged your work unfairly. Peer assessment enables you to critically formulate and structure your information to get the best out of your work.

3.4   Self-assessment

Self-assessment in many ways is similar to peer assessment with the additional advantage that you can see how you are progressing through a module or programme of study and reflect upon your development. This is important as it is possible to identify strengths in your work, and resources needed to address any weaknesses. Students studying with you at university are drawn from a wide range of areas and backgrounds within society and are motivated to study for different reasons. Some of you who want to study Criminology may do so because you feel that you already know something about crime and society and have an interest in or are curious about crime. Others engage with the subject because they hope to take up posts within the criminological professions, such as the Police, the Courts, **Probation** Service or Prison Service while others may already be employed in these or other criminal justice agencies. The current tendency towards lifelong learning is often seen as a two-way process that marries together theoretical concerns on academic courses with practical skills in the market place, and gaining qualifications in a particular subject area is necessary for career advancement. Employers also increasingly require their employees to update their skills to keep pace with fast-changing economic circumstances that take place within a global market. Taking these factors into account together with the trend towards increasing professionalisation of many areas of work, academic study becomes a significant part of the professional ethos.

Activity 3a

Think about why you want to study Criminology and what motivates you towards this subject area.

List professional and personal reasons.

This will enable you to be selective about the modules and the assessment that you do. It can help you to develop specialised knowledge to support your future career aspirations.

As students you will bring different life experiences to your studies and it is recognised that each individual student has different limitations and strengths. Practising self-assessment will enable you to locate any specific difficulties you may be experiencing and to think about what you might need to do to resolve them so that you can progress and achieve your potential. As well as the above, self-assessment can take many forms including checking your skills in accessing library resources, organising your work environment and schedules to meet deadlines, reflecting upon how you meet your targets, reviewing your writing and presentational skills and considering what might be done to improve your work.

Activity 3b

Think about the last examination, essay or presentation that you have carried out. Consider how you organised your research and revision and critically assess what you would need to do to improve it.

Studying Criminology at university encourages you to see crime and criminal behaviour from a variety of perspectives that will broaden and deepen your knowledge of the topic. It gives you a better grasp of the subject area and enables you to critically evaluate and question taken-for-granted assumptions about crime and society. Assessment is necessary to measure your learning and self-assessment can play a key part in developing your technical skills, improving your critical thinking and increasing your knowledge and understanding of Criminology.

3.5   Learning outcomes

Learning outcomes are the intended skills and knowledge that you will acquire by studying a particular module. The following shows some examples of learning outcomes from Criminology modules.

A Criminal Justice module might state that on completion, 'you will be able to':

- understand major themes in criminological policy

- recognise the social factors affecting the processes underlying the formulation and implementation of policing, prosecution, sentencing, penal and victim policies

- comprehend rationales underlying different custodial and non-custodial sanctions and recognise the character and effects of their implementation

- critically evaluate criminal justice policy and practice

- appreciate the social, political and ethical issues arising in criminal justice policy

- develop skills in communication, interpretation of data, problem solving, decision making and critical analysis.

For a 'Race', Crime and Criminal Justice module, expected learning outcomes might state that upon completion of the module, 'you should be able to':

- familiarise yourself with the range of sociological and criminological theories about 'race' and crime

- understand and distinguish different racisms (for example, direct, indirect, institutional) and their operation within society and the criminal justice system

- comprehend the ways that racism and racial discrimination operate within the criminal justice system

- analyse critically the different interpretations placed upon the relation between 'race' and crime

- appreciate the contribution of different theories of racism, ethnicity and identity towards criminological understanding of victimisation, offending and criminal justice.

These are provided only as an indication of the types of learning outcomes, and students should not expect to see exactly the same outcomes in similar modules throughout all universities. It is important to note that when you are being assessed, these outcomes must be evident in the work you have done and it is how well you demonstrate that you have met them that contributes to your final grade.

3.6 Marking criteria

This is an indication of the typical criteria that assessors will be using when they mark your work. The example below is based on the criteria used at the University of Teesside and may vary somewhat in other institutions but the general principles will be broadly similar. External examiners are employed to ensure that the criteria used in different institutions are compatible to that in other similar institutions. As you can see 40% indicates a pass grade while 70% indicates very good work indeed.

AN EXAMPLE OF MARKING CRITERIA

70%–100%
Points are made clearly and concisely, always substantiated by appropriate use of source material. There is evidence of a sound ability to interrelate critically theories with examples from practice. The work contains coherent arguments with some evidence of original thought. Presentation is excellent.
60%–69%
Very good presentation with an emerging ability to apply knowledge critically to practice. Appropriate evidence, good use of source material, which supports most points clearly. Content is wholly relevant, within a fluent coherent structure. Critical reflection could be developed further.
50%–59%
There is demonstration of a sound knowledge base, but limited critical and practical application of concepts and ideas. Content is largely relevant although points may not always be clear, and structure may lack coherence. Use of source material to illustrate points is generally adequate but may be lacking in some instances. Contains some critical reflection. The presentation is of a good standard, but with minor errors in grammar and spelling.
40%–49%
Adequate presentation with some errors. The work is descriptive but relevant, with clear evidence of knowledge and understanding. There is evidence of some reading and there is limited critical reflection. Links to practice are made, although arguments are often lacking in coherence and may be unsubstantiated by relevant source material.
39% and below—fail
Poorly structured, incoherent and wholly descriptive work. Limited evidence of appropriate reading, and no evidence of critical thought. Referencing poor or missing.

3.7 Choosing modules to study

At various stages in your programme you may be given the opportunity to choose which modules you want to study (option modules are discussed in Chapter 2) that have a particular form of assessment criteria. Quite often these are modules that do not have examinations or presentations and instead use essays or other forms of in-course assessment. However, a range of assessments is likely to be used throughout your degree programme, and examinations and presentations as well as essays give you the opportunity to show your knowledge and understanding of the subject area. Depending on your programme of study, you will be required to take a number of core modules; that is, some modules will be deemed essential to obtaining a subject specific degree. As previously mentioned you will also be able to choose some from a range of other modules that you find interesting, and it is important that you base your choice on the content of the module rather than on what the assessment is. If you are interested in a specific topic area you will be motivated towards learning about it and gaining a thorough knowledge of the subject, whereas if you are taking modules based on what the assessment is there is a greater likelihood that you will become bored and disinterested quite quickly. It may, therefore, become a struggle for you to engage fully with the content and gain information that is necessary for you to pass the module.

A variety of assessment is also very useful for life after graduation. You may want to take up further study and research within Higher Education, such as a Masters degree programme, and the various techniques learned during your undergraduate studies will give you the confidence to pursue this. As we have discussed earlier in this chapter, within many work environments there is the requirement to write reports, make presentations and utilise your knowledge in those situations when you have to 'think on your feet'. This is discussed in more depth in Chapter 11.

3.8 Relative weighting of assessments

You need to make sure that you are aware of the weighting related to the assignment that you are undertaking for assessment. For example, you may be asked to answer an essay question and are informed that this will be accorded 50% of the overall module mark, an essay plan will make up 10% of the overall mark, and a presentation will make up the remaining 40% of the module. Giving some consideration to the weighting of these specific elements is useful in terms of the time and effort you spend on these tasks. In this example the essay plan is only worth 10% of the overall mark and, therefore, you will not put the same amount of time or effort into completing this as you would with either the presentation or the essay itself.

You must always check when each component of the assessment needs to be submitted, since they may have different deadlines, and ensure that you know where you need to submit these. In the above example it could be that the essay plan would have to be submitted in advance of the essay and this may be followed by the date for the presentation. Even though within this example there is no examination, these dates and details regarding the assessment ought to be treated in exactly the same way as you would approach an examination, as they are just as important.

3.9 Assessment range

Assessment takes place in various ways across the whole spectrum of university degree courses and sourcing knowledge and information necessary to meet assessment criteria, particularly within Criminology, takes many forms. Assessment does not necessarily depend upon one person's contribution but can depend upon the input of other individuals; for example, seminar presentations may involve co-presenters and peer group grading. Formal lectures are the means by which lecturers will provide you with key points about particular aspects of Criminology; however, it is within seminar groups and workshops that discussion and often critical debate takes place. Lecturers use various techniques and tools to introduce students to criminological theories, policies and practices and, although seminar activities are not graded, taking part in these activities is a productive aid towards assessment. For example, you may be asked to read several articles about the growing prison population and to highlight the similarities and differences between each writer's perspectives. You would then critically assess what you have learned and this could contribute towards a framework for an essay title or examination question.

Activity 3c

Gather articles or chapters that focus upon the reasons why penal policy in recent years has become more punitive in respect of offending behaviour. Critically evaluate competing values.

Focusing upon and researching a specific topic area and summarising it for a seminar discussion group is another way of gathering information that can be used in your assessment.

Activity 3d

Read a chapter on the way crime is reported in the media and summarise the key points that you think could provide the basis for discussion in a seminar group (see for example Reiner, R. (2002) 'Media made criminality: the representations of crime in the mass media', in Maguire, Morgan and Reiner (2002: 376–416)).

Taking part in seminars and workshops is an essential part of your university career and whether you are a mature student or entering university straight from 'A' levels, the variety of skills that you develop and learn during seminars is invaluable in respect of your progress as an undergraduate. Whether they are graded or not, different types of assessment together with seminar activities and discussion groups are excellent ways of gaining confidence and building up a knowledge base that can be drawn upon across the range of assessments.

3.10 Summary

- This chapter has shown you why assessment takes place in Higher Education.

- The different types of assessment that you might undertake on a Criminology degree programme have been identified.

- Expected learning outcomes and marking criteria have been introduced to give you an awareness of expectations in Higher Education degree programmes.

- The usefulness of taking part in various types of assessment during your degree has been discussed and the relevance and benefits of these skills in future study or employment have been highlighted.

4 Presentation Skills

CHAPTER OVERVIEW

By the end of this chapter you should be familiar with:

➢ undertaking presentations
➢ report writing
➢ portfolio building.

Oral communication skills such as formal presentations are used as part of the learning and teaching process more and more often by having assessed presentations within Higher Education, but also for job interviews you are asked more often to prepare and present a short presentation on a related subject. Such communication skills are considered very important, since they illustrate how clearly you can communicate information to others and, also, they demonstrate your organisational skills in terms of the coherence and construction you give to the presentation. These skills are clearly highly regarded well beyond the end of a degree course and are the kinds of interpersonal skills that employers often look for.

4.1 Presentations

When asked to deliver a presentation you will need to consider:

- the audience

- the venue

- timing

- visual aids and props

- the type of presentation.

THE AUDIENCE

In relation to this, think about whom you will be presenting to and how you can make your subject matter interesting to that audience. Perhaps think about a question or two to ask the audience at the end, or attempt to get the audience to engage with a discussion of some of the points you cover within your presentation. Eye contact with the audience is very important. You are communicating your information to the audience so try to engage with the audience by using some eye contact. Make sure that your audience can see you, can see the props you are using, such as PowerPoint, an Overhead Projector (OHP) and any handouts, and can hear you clearly. Be sure to pitch your presentation to suit the audience you are presenting to and pace your presentation within the time frame without rushing through the topic.

KEY TIPS

The purpose of presentations is not to provide a list of facts and figures. These could be referred to and fuller lists provided in a handout at the end of the presentation.

Presentations should communicate information and stimulate discussion and debate. Questions from the audience at the end of your presentation will show whether your discussion was thought provoking.

THE VENUE

This is important in the sense that if you have prepared a PowerPoint presentation you will need to know if the room is suitable and consider how many people are likely to be attending. You may also need to consider the seating layout and this may differ if it is a formal or informal presentation that you are undertaking. It is always a good idea to take time to visit the venue before you are due to present there.

> **KEY TIP**
>
> Do make sure you gain some eye contact with the audience while you are presenting. It is very off-putting to attend a presentation where the presenter avoids eye contact and this may lose you marks.

TIMING

You will usually be given a time limit for your presentation. Make sure your presentation conforms to the allotted time slot that you have been given – if your presentation is too long and goes on too long you may lose marks and lose the attention of your audience too. This would also have a knock-on effect for other presenters if more than one presentation is taking place. Pace yourself, speak clearly and try not to speak in a monotone.

> **KEY TIP**
>
> Practise the timing of your presentation complete with pauses and time allowed for prop use in a dry-run session for friends, family, pets or simply for yourself, but do this aloud so that you can think about your voice and the pace of the presentation as well as the volume you will need to speak at.

VISUAL AIDS AND PROPS

You will need to consider which props may be useful in order to provide a good presentation. When making a presentation you may find it useful to use the Overhead Projector (OHP) or PowerPoint so that the main structure of the presentation can be displayed to everyone while you talk through your material – pointing at important points. Make sure that you use a large font size when printing onto acetates to be projected so that your audience can clearly see the information, and do not overload each page – four or five points are usually sufficient to talk around on each of the acetates. This also has the added advantage of taking the spotlight away from the presenter so that the audience is looking at whatever is being projected as well as the presenter.

Handouts for the audience can also be useful in setting out important points or definitions. Remember that a clear structure is as important with a presentation as it would be for other assessments and this can be followed by signposting the structure along with you uncovering the OHP point by point then discussing it.

KEY TIPS

Try to avoid telling jokes at the start of a presentation – this will not be expected. Provide some structure so that the audience are aware of what your presentation will cover.

Handouts should not contain all your presentation notes (the key areas are usually sufficient) or the audience will read the notes and not look at you!

Make sure you have the tools necessary for your presentation such as flipcharts and pens, handouts and acetate slides.

THE TYPE OF PRESENTATION

Different types of presentation will require different uses of these tips and hints on how to undertake a presentation. On the issue of the content of presentations and how to structure them, we would suggest using the hints and tips we have covered in essay writing but, rather than simply having a written finished product to submit, a presentation would include the delivery of those ideas – your speech and the speech of those you co-present with, for instance. If you are undertaking a group presentation it is vital to divide the research, writing and delivery, and then, just as important, that you do several practice sessions since the timing becomes even more significant when there are more people involved.

Try to think about some questions that can be used to get the topic discussed after the presentation and be ready to answer questions from your audience.

KEY TIPS

Check how long your presentation is and be realistic about the number of presenters. Six people undertaking a 20-minute presentation would have about three minutes each, which is hardly enough time to make a worthwhile point.

Don't rush through it to try to get it finished as soon as possible – this will leave your audience frustrated since they will not be able to follow or understand what you are saying.

Avoid reading your presentation from a sheet – you are presenting and should use your voice, body and language to get your points across to the audience.

Don't overload your audience with too much information in your presentation or on your visual aids – keep to the most important points and keep it clear and brief.

Sometimes you will be asked to submit a hard copy of your presentation to your tutor so make sure you have your own copy with notes jotted on and keep a clean copy of the presentation and any handouts for your tutor.

4.2   Report writing

Reports are written for a variety of purposes and usually have recommendations in order for some action to be taken. These are generally concise and incorporate, for example, sub-headings, lists, bullet points, charts and graphs.

Reports are also different from essays in terms of the layout, since essays have a flowing text and do not usually include bullet points and sub-headings but rely upon sentences and paragraphs. Rather than culminating in recommendations, essays provide informed debate, critical analysis and understanding. Reports can be a valuable and quick reference on many subjects but they must be clear and concise – check that any tables or diagrams you use illustrate findings easily and are easy to read. Remember to put a key in where necessary, or an explanation so that this is easy to read at a glance. If you are unsure about how to display findings have a look at how other reports have displayed their conclusions.

KEY TIP

If you have never seen a report then go to your library and have a look at the layout and structure of an example. It is useful to see how other similar work has been presented.

The following shows you how to structure a report and what the content should include:

- Planning and researching for your report is the same as it is for essay writing and examination preparation.

- You will need to read the brief for the report and check that you understand what it is asking you to do. If you are unsure, check meaning with your tutor.

- Use sections with headings and sub-headings throughout your report and number and give titles and sources for all graphs, tables, figures and diagrams. Make sure you refer to these within the text.

- Your report should therefore include:

  1 Title (make sure the title is clear)

  2 Contents (the contents page will allow the reader to dip in and out of the report)

  3 Executive summary/abstract (this should be a summary of the whole report including the recommendations in no more than two paragraphs so that the reader should be able to tell what is in the whole report by just reading this)

  4 Introduction (explain the purpose of the report, indicate the content, and highlight the main issues/topics and how you will deal with them)

  5 Main body of the report (this part contains the majority of material relevant to your assignment — it should be presented clearly and include an explanation of the research methods used and findings. Use the tools of report writing — bullet points, sub-headings and graphics to arrange your material clearly)

  6 Conclusion (you should summarise the main points of your work without repeating what you said previously and draw a definite justified conclusion)

  7 Recommendations and implications (list your recommendations)

8 Bibliography (correctly set out the sources you have used to inform your report)

9 Appendices (optional – appendices are exclusive of the word limit, but use them sparingly to give supporting information and remember to refer to them in the report).

KEY TIPS

Do not forget to edit your final draft by checking typing, spelling, grammar and punctuation. Write your contents page last then number the pages so that any changes will not mess up your contents page.

Add graphs last, otherwise they tend to jump off pages and cause trouble with the text on the pages.

Leave plenty of time for word-processing your report and displaying findings, and build in some time to allow you some flexibility if you encounter any problems.

Activity 4a

Go to http://www.homeoffice.gov.uk/rds/pdfs04/dpr26.pdf on the Internet to have a look at one of the reports that you will find there. Identify the key differences between a report and an essay or a textbook. Make a note of the key differences.

4.3 Portfolio building

A portfolio is simply a collection of work submitted together – different elements would be placed in the same folder and submitted as a portfolio of work. Practical work can be submitted in a portfolio alongside theoretical work in order to evidence these skills. Portfolios are used more and more often as assessments for study skills and research methods since they are a way of illustrating that students have progressed to achieve certain standards in their work. An example of a portfolio for a study skills module might be for students to submit an essay plan, notes for their essay, the actual essay with bibliography and a short reflection on the process of essay writing.

With a portfolio you need to:

- design a contents page for your portfolio — this will make finding different work within it easier and will provide clarity

- make sure each element within the portfolio has a clear title on it

- make sure each piece of work in the portfolio is in the right order

- check spelling, grammar and punctuation.

KEY TIP

Use dividers within your folder to separate the sections and label the sub-sections since this will provide clarity and structure to the portfolio of work.

4.4 Summary

- This chapter has illustrated some useful tips and guidelines when faced with undertaking a presentation, building a portfolio or writing a report for the first time.

- It is important to remember that, as with essay writing skills, the skills used for these purposes can be used in different contexts and such inter-personal skills are vital in many different occupations in order to provide clarity, to illustrate clearly and coherently your views or to explain meanings within the world of work as well as within education. Getting into the habit of following these guidelines is therefore useful for the whole of your degree programme and beyond!

5 Examination Assessment

CHAPTER OVERVIEW

By the end of this chapter you should be familiar with:

➤ the purpose of examinations
➤ different types of examination
➤ ways of dealing with pressure
➤ strategies and techniques to enable you prepare yourself for examinations.

Examinations can cause anxiety for many of you and, as indicated in Chapter 3, some of you may choose modules or units of learning that do not have examinations as an assessment criterion. However, the apprehension that you might feel about taking examinations can be reduced and viewed in a more positive way. As stated in Chapter 3, as students you enter university with varying levels of skill and expertise and bring with you a diverse range of life experiences. The students within any cohort will study in different ways and you will probably tackle research and writing using a variety of approaches that reflect your previous learning experiences; however, as with other assessment methods, examinations have a number of golden rules to follow if you are to achieve the desired outcome and be successful. This chapter will begin by considering the purpose of examinations as a form of assessment and the different types of examinations that you might encounter within the academic institution. Ways of planning and preparing for examinations and some techniques to support your revision will be highlighted as well as strategies to strengthen your performance within the examination and this will hopefully minimise any anxiety you feel. Preparation and planning for examinations is key and familiarity with the rules, regulations and protocols required during the examination process will make you feel more confident and help you to perform successfully.

5.1 Why examinations?

Examinations are often perceived by students to be an unnecessary, stressful experience and many of you will share this view. It is often evident to lecturers that tension, anxiety and fear loom large when examinations are due. Those of you who are mature students and have been away from education for some time may feel most disconcerted by examinations since it may well be the first mention of examinations since you left school. Whether you are a mature student or a student who has come to Higher Education immediately after completing your 'A' levels, few of you will rejoice at the idea of having to undertake an examination. More recently there has been a move away from unseen examinations with undergraduate degrees utilising different assessment methods, such as seen examinations or take home examinations; whatever form they take, however, examinations do have many advantages and continue to be a useful assessment tool in a number of ways. Assessment by examination is often seen to be a way of testing whether you have gained the necessary knowledge about a specific subject; however, this is rather a narrow view. It can also be seen as a very useful tool for both you, the student, and the lecturer. For example, the examination process gives you the chance to test your study skills and show your breadth of knowledge in a time-constrained environment. Focusing upon a course of study for examination purposes gives you the opportunity to get an overall awareness and knowledge of the module that you are studying. This is important as Criminology programmes can relate to a wide and varied range of employment opportunities that increasingly require subject specific degrees. Examinations demonstrate to potential employers and professional bodies that you can meet the required academic standards that underpin courses in Higher Education. Being able to apply your knowledge under pressured and stressful situations is an extremely useful skill that is valued by many organisations and professions that rely on their employees being able to make decisions under pressure. Feeling pressure can, therefore, be a very positive and creative force and managing your performance in examination situations can be seen as a great personal achievement.

Examinations also enable lecturers to assess whether you have understood the course content. The questions set by lecturers are usually directly focused on key topic areas and the outcomes required for successful completion of the module. While essay writing often focuses upon one particular aspect of a subject area in some depth, examinations allow you to be tested over a wider variety of issues across several aspects of the module and lecturers can assess whether you have a broader range of knowledge and understanding of key themes, perspectives and theories.

5.2 Types of examination

For some modules you may take an examination at the end of the course, while for others the examination may be in course to test your knowledge during a particular part of the module. As mentioned earlier, examinations can take several forms: they may be a seen paper, a take home examination or an unseen examination.

- *A seen paper* is an examination paper that would be given to you at some time before the examination takes place. This could be two days or two weeks before the examination and this time is given so that you can prepare your answers beforehand to recall at the time of the examination. You will not usually be allowed to take notes into the examination room in this instance.

- *A take home examination paper* is exactly that; you would take home the examination paper to research and complete at home within a given time scale. In this instance you have access to notes and other resources when answering the questions but will have no knowledge of the questions until you receive the examination paper.

- *Unseen examinations* are probably more familiar to you and it is likely that it is this type of examination you would definitely be required to do at some stage during your degree programme. An unseen examination means that you do not have sight of the examination paper until the examination actually takes place, although you may be given an indication of the broad areas that the questions will focus on during lectures or seminars.

Whichever type of examination you are undertaking you may be required to answer some or all of the questions. For example, it might be the case that you will have a choice of questions in seen and unseen papers but that a take home examination may specify that all questions should be answered. It is important to read the instructions as it is essential that you know exactly what is required of you. The examiner will only mark what is presented and if you answer too few questions you will significantly reduce your chances of gaining a high grade. Equally, if you answer too many questions you may not have given yourself the chance to do well as not all questions will be marked and you will not have used your time effectively.

> **KEY TIP**
>
> When you have an either/or question, make sure you only answer one of the two choices, as the examiner will only give a grade for one.

5.3 Planning your examination preparation

A strategy for revision and time management is essential. When you have several different modules and subject areas to submit assessed work for it is important that you plan in advance and prioritise your workload.

> **KEY TIPS**
>
> Make a list and diarise which examination takes place first and locate these within the wider picture of assignment submission. This enables you to structure and timetable your research accordingly.
>
> Allow some flexibility into your preparation programme so that you can deal with any unforeseen circumstances.

Make sure you know the date, time, place and duration of each examination you are taking and clearly diarise this. If you are unfamiliar with the location of the examination hall or room it is always a good idea to go and find where it is and how to get there. Suddenly finding that you have lost your way and may be arriving late for your examination will only increase your anxiety. Examinations frequently last for three hours, but some may be one or two hours long. You need to know what the duration of the examination is so that you can revise effectively. It would be a waste of your valuable time to revise, say, four questions for a three-hour examination to find that it is only two hours long when you get there as you would probably not have sufficient time to reproduce your answers as fully as you would like.

5.4 Organising your examination revision

A timetable for revision is useful and should be drawn up well in advance of the examination date. You should also include time for family and socialising in your timetable and try to organise domestic life, paid work and any family commitments around your revision schedule. These are factors that you also need to plan for and organise for the day of the examination itself so that you can focus your attention on the task before you. You need to prepare yourself to produce knowledgeable, well structured answers in the examination and the only way that you will be able to do this satisfactorily is to revise effectively.

Activity 5a

Construct a timetable for examination revision. List topic areas to be revised and resources needed. Try to set realistic targets and remember to include some flexibility into your schedule to allow for anything that occurs unexpectedly.

5.5 Gathering information

When you begin to prepare for an examination you will use a range of resources. These will include your own lecture notes, any notes provided by the lecturer, seminar readings, your notes from your reading, the module guide and any other literature and articles relevant to the course that you have access to. Make sure you do the research around the areas that you are interested in answering questions on. If the lecturer's notes are available on your university intranet site, or other electronic learning environment, then print these to begin your research and then search for further resources to develop this in the library. It is a good idea to use a range of source material rather than relying on one or two sources; for example, extensive or sole use of websites is not recommended as many sites may be dated and material may be information rich but knowledge poor. By this we mean that there may a lot of information but the depth of analysis and the evidence base for sound academic work may not be available. Literature that contains in-depth knowledge, discussion and critical assessment of significant criminological points is much more useful and shows breadth and depth of research to the person marking your examination. This is more likely to lead to your answers being well structured and relevant rather than descriptive and unbalanced.

KEY TIP

Make sure you attend any examination revision sessions that might take place at the end of the module. These sessions can be invaluable as a revision aid as you will revisit crucial key areas of the module.

5.6 Summarising information

Once you have gathered your resources together you should begin to make your notes. If you are preparing for a three-hour examination where you have to write three essay-style answers then you will need to get three sets of notes ready. Make sure that you note down any authors whose work you have used. Remember that the information needed to write an answer to examination questions should be broad and take the form of key points, ideas and theories discussed on the module. Using the literature, read the relevant topic areas and summarise the key points into short paragraphs that make sense to you so that you can convey that understanding to the person marking your examination paper. Begin writing essay plans in the same way that you would if you were writing an essay (see Chapter 8) but just use key words and sentences in it instead of producing a full discussion. In this way you can remember, say, a ten-point plan to go into the examination with. This will help to give some structure to your work under examination conditions and should ensure that you have some cues to develop your answers from. Remember to include the names of authors and theorists that you can cite within your examination and what they have to say on certain issues. This will illustrate to the reader of your examination paper that you have undertaken some relevant research in preparation for the examination. In an unseen examination you are unlikely to be required to produce bibliographic details but you will be expected to attribute work to the relevant sources.

Activity 5b

Read a chapter in one of your Criminology texts and summarise it in ten bullet points. This will help you to think about key terms, phrases and debates as preparation for your examination.

Look at your notes with a critical eye and try to condense them. Restructuring your knowledge in this way to summarise key points for examination purposes is a useful skill, and it is often helpful to rewrite your notes from time to time as this helps you to memorise the key themes of your information. From these key points you will be able to construct an essay plan and this reworking and rewording of your research notes into a coherent form reinforces your knowledge of the topic and provides structure to your examination answer.

The following example of an examination plan provides ten short paragraphs to show how you might answer a question about the competing histories of the emergence of the 'new police' in Britain during the early nineteenth century. You will probably need to adapt this to suit your own style but the basic principles will be helpful to you when taking an examination.

EXAMPLE OF AN EXAMINATION PLAN

Question: The emergence of the new police in England and Wales has been interpreted in divergent ways. Critically evaluate two contrasting perspectives.

- There are two competing theories, orthodox and revisionist. The two share some similarities but also have some significant differences. When considering the two perspectives chronological events are not disputed; however, the underlying ideology and justification for the creation of a police force is subject to debate.

- Orthodox view — police introduced as a result of logical progression from existing 'police' systems, for example, nightwatchmen, tithing men, parish constables (Emsley in McLaughlin and Muncie, 1996).

- As social change occurred masses of starving, homeless people posed a menace to 'authority' and fear of the mob necessitated policing to ensure social order.

- Revisionists see the introduction of the police as a consequence of middle-class fear relating to public order and rising crime.

- Police were a force to protect the minority from the majority effectively enforcing the rule of law as prescribed by the state.

- Mass meetings protesting against poverty became 'increasingly stigmatised as criminal' and state legislation increasingly supported and created the emergence of a new police system.

- Which of these appears to be the most logical given the development of policing over the nineteenth and twentieth centuries?

- Social change relates to increase in urban population creating large numbers of unoccupied people. Distinction made between honest people who would work hard and those who preferred idleness. Consequently there was a need for more police who were better organised.

- Equally there was a rise in crime that was attributed to the unruly working classes or the 'dangerous classes'.

- Fear of crime led to demands for greater protection from the middle classes.

The essay plan outlined above provides ten key points that focus upon the specific question and is structured to introduce the topic, provide a critical evaluation of two competing perspectives and a summary that draws the threads of the discussion together. As mentioned above, some of you would probably approach this question differently and would summarise your research more briefly or more elaborately to suit your own learning style. The significant point here is to find a revision style that suits you and would enable you to provide good answers that clearly illustrate your understanding of the question.

Activity 5c

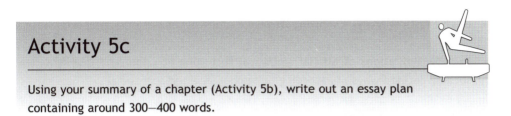

Using your summary of a chapter (Activity 5b), write out an essay plan containing around 300–400 words.

This technique of condensing key points into a structured form also enables you to construct an argument that is an essential requirement of any essay and examination response. All essays start with a question or statement and your response should produce a relevant discussion that leads to a conclusion.

5.7   Memory strategies

There are various strategies that you can use to revise and remember the information that is needed to answer examination questions. One of these memory aids is the use of acronyms. This is where a word is formed from the initial letters of other words, for example, anti-social behaviour order (asbo). Another method is where each initial letter of the word or phrase prompts another word or sentence. You may be familiar with the rhyme, 'Richard Of York Gave Battle In Vain', which is used to remember the colours of the rainbow – Red, Orange, Yellow, Green, Indigo and Violet. An example that is more relevant to Criminology could be to use the words 'crime survey' to help you organise an essay plan that critically evaluates 'how and why crime statistics are gathered'. The discussion could be structured as follows:

- to introduce generalised perceptions of rising crime

- to show how official crime statistics are gathered

- to consider political and social concern about crime via policy measures and media reports

- to describe and discuss newer developments and methods of categorisation (i.e. British Crime Survey) that appear to offer a more valid picture of crime

- to give critical analysis showing strengths and weaknesses of new approaches

- to show limitations of official statistical recording of crime and highlight new socially defined ways of recording behaviour deemed anti-social

- to summarise by reflecting upon what it is that we are measuring, who measures it and why.

Using the words 'crime survey' as mentioned above we can construct the following example of an acronym:

C = Continuing rise in crime figures

R = Reporting and recording practices

I = Institutionally produced criminal statistics

M = Media, moral panics, the public and politicians

E = Emerging trends, new sources

S = Surveys

U = Usefulness or not of questionnaires

R = Reality of rapid crime increases, the radical critique

V = Victims, victimisation and risk

E = End of official statistics, rise of recording socially defined incidents

Y = Yardstick — measuring what, by whom and why

This brief plan shows that by using a simple acronym a whole range of significant key features of this topic area can be prompted and brought into memory to be recalled during your examination.

Activity 5d

Consider an examination question or a statement and put together your own acronym that would prompt you to produce a structured response within an examination situation.

There are other useful ways of prompting memory for examination purposes. Rather than writing material out in a linear fashion, diagrams can be used to stimulate memory. For example, you may find spider grams useful. This is where the main topic is written in the centre of a page and themes connected to that topic are radiated out from that central image. You can then add more branches or arrows to link ideas and convey more key words and images to create a one-page revision tool. You may wish to use different colours to highlight specific blocks or main ideas. Mind maps such as these are seen to be extremely powerful revision tools as you use both sides of the brain to construct the diagram. The creative aspects of this means that you should remember the information more clearly than if you just had to memorise lines of words.

Another way of remembering key ideas is to take a journey around a place that is familiar to you, for example, a walk around the area that you live in or in your home. As your journey takes you past familiar streets, shops or signs you can memorise these in a way that connects to particular parts of your examination plan. In the examination you simply recall the route you have taken to remember specific points. Similarly, a virtual walk through the living room, dining room and so on can connect material together.

Whatever method of revision you use, the main purpose should be to find ways of recalling meaningful information when you take the examination. You can use this imagery to make associations between relevant key principles or use strategies that utilise memory by rote, such as reading over the same chapter and your revision notes. The main point here is not to be anxious about revision but to find ways of revising that are suitable for you. If one method does not work for you, persevere and try others until you find one that does. You will not necessarily prepare for an examination in the same way as your friends, family or other students whom you study with. The key here is: 'If it works for you, then it is OK.'

5.8   Past papers as a revision aid

Using past examination papers as a revision tool is extremely useful in several ways. For example, you will gain a good idea of the way in which examination papers are set out and the types of questions that may be asked. You can also practise answering them! You will be able to focus upon the question to determine what it asks of you and construct appropriate essay plans to answer the question. A dry run of the examination using past questions is a useful revision exercise on the subjects that you are interested in answering. Even though the questions change it is often the case that the key themes remain similar. It is a good idea to time yourself to see how long it takes you to write your response. If you are required to answer four questions over a three-hour period you will need to take about 45 minutes per question including thinking about your response. Practise writing your answer in the given time and then check your work to make sure that you have answered what the question requires in the way that you intended.

KEY TIP

Check with your tutor where copies of past examination papers are held, for example, the library or other resource centre, and go and get some to have a look at the way questions have been asked on past papers.

Activity 5e

Look at the following questions and break them down into the parts that you think need to be focused upon. Write out a brief essay plan for one of them so that it clearly identifies and addresses the relevant key points.

1 Policing in Britain has changed significantly over the past two decades. Critically analyse relevant key developments.
2 Why has the prison population in Britain increased so much during the past ten years and is the greater use of imprisonment likely to reduce crime?
3 Non-custodial penalties are not as effective as custodial sentences. Discuss.
4 All are equal before the law. Critically evaluate this statement with reference to gender, ethnicity and social class.

Think carefully about your answers. Analyse the question, highlight key ideas, perspectives, theories and approaches that may be required and practise writing out your response.

5.9 Examination strategies

Now that you have carried out your revision, made essay plans and practised answering questions on past papers, it is necessary to think about how you deal with the examination itself. There are some strategies that will help you to get the best out of your performance and if you think about them you can reduce the anxiety that surrounds this type of assessment.

UNDERSTAND AND ANALYSE THE QUESTION

You must focus upon what the question asks you to do. As with essay writing the question should be carefully scrutinised to ascertain what needs to be answered. Read the question carefully and identify the parts of the question that require an answer. No matter how well written or well presented your answer is, if it fails to address the question, you are likely to lose marks.

KEY TIPS

Focus upon the question, and answer only what is required.

Do not write everything you know about the topic, as you will only be graded on the information that answers the question.

WATCH YOUR TIME AND MAKE A FRAMEWORK

Be aware of the length of time you will have available to answer each question and time yourself accordingly. It is often a good idea to do your 'best' question first as this should give you confidence to address other questions you may not feel so sure about.

You may want to spend a few minutes mapping out your responses to particular questions so that you have a framework for each question you intend answering. You will usually be able to use the inside cover of your examination booklet for this or there may be an indication on the front cover of where you can do your notes. You should normally score through any notes you do not want to be included in your answer.

If you feel your revision might not adequately address a question you are interested in, spend a moment or two jotting down any connections you can think of but do not waste time agonising over it. Far better to get on with those you can answer and then spend time later looking at a potentially 'weak' one.

Here are some other helpful points:

- If you are writing your final answer and find that you are running out of time to complete it, you could write the rest of your answer in bullet points that would indicate to the marker how you intended answering the question. Of course, you will not be marked in the usual way for this but you may gain a valuable point or two if it clearly shows your understanding of key areas.

- Do not leave the examination early; it is a better idea to go over your answers and check them for focus, quality, spelling, argument, sense and so on.

- Try not to panic too much! If you have carried out your revision you should be able to write good answers.

- Do not carry out a post-mortem afterwards; this rarely achieves anything but worrying you about what you should or should not have written. It is unlikely that you and your fellow students will have all answered the questions in the same way, but all may be worthy of a good mark.

- Wait until you receive your results and then think about whether you need to ask for feedback to address any anxiety you may have about your performance.

- Remember to structure your answer. The section below gives some tips about expected writing style.

5.10 Writing style

The usual writing style for examination purposes is slightly different to that required for essays. The expectation is that you will have a broad range of knowledge and you will be able to apply key points to the question. Lecturers marking your examination script will be looking for you to show that you have focused on the question and that your answer demonstrates that you have carried out relevant research into the subject area. It is breadth of knowledge of appropriate significant key theories, perspectives and authors that you will need to present to indicate your understanding of the topic. This is why the section on condensing information into concise bullet points is so useful as each point can be turned into a paragraph or several paragraphs. Remember that you need to have a structure that flows, so, to present your answer well, you should: introduce the question; discuss in the middle section what the main purpose of the question is, paying particular attention to the criteria specified (that is, are you being asked to critically evaluate, discuss or assess?); and then summarise the threads of the argument in a conclusion. As mentioned above, writing your best answer first can often be useful as this gives you more confidence to tackle other questions; however, do remember to keep an eye on the time you spend on each answer. Writing a lengthy first best one at the expense of finishing all questions required is not a good idea and is unlikely to gain you a pass grade no matter how good the answer! Remember the marks available for each question will be the same so a brilliant answer that gets 80% cannot fully compensate for a weak answer that only gets 20%.

5.11 Using references

You should always cite authors and sources of work that you are using, even during an examination. You should not write a bibliography, as it is not expected. The intention should be to show that you have knowledge of and can refer particular theories, perspectives or subject areas to specific authors. Therefore, you should cite the author and the publication date in the text. This sounds more difficult than it perhaps is as most questions will focus specifically upon one topic area that has been covered during the module and some of the relevant writers and theories will usually be familiar to you. You will probably be surprised how much and how many authors and theorists you actually do know!

Some examples include:

Winlow, S. (2001) – night-time economy

Walklate, S. (1989) – victimology

Pearson, G. (1983) – youth crime and historical amnesia

Farrington, D. (2002) – criminal careers

Murray, C. (1990) – underclass.

5.12 Protocols to be observed during examinations

You need to familiarise yourself with the protocols that are expected during the examination process. There are rules and regulations governing academic and intellectual behaviour that you may not have been familiar with before you entered university, such as referencing work, plagiarism and collusion, and these can usually be found on university websites and in student handbooks and module guides. Putting these into practice is essential. However, there are also a number of other points that can be raised and discussed that should make you feel even more comfortable about going into the examination environment! The following information and hints should be useful to you both in your preparation for the examination and when sitting the examination itself:

- You should already have noted the place, time and duration of the examination. When the examination is due to take place it is a good idea to arrive in good time as before the examination begins you will be directed to leave your bag or folders to one side of the room and you will have to seat yourself and fill out some forms. *It is usually expected that you arrive at least ten minutes before the examination is due to take place.* Make sure that you bring your student number with you! Many universities now require you to bring your student identification with you; failure to do so may result in your exclusion from that examination.

- When you enter the examination room the examination paper will be on the desk in front of you. You must not turn this paper over until you are directed to do so by the examination invigilator.

- Once you have been instructed to complete registration cards and put your name on the answer paper, you should not communicate with other students. This remains the case throughout the duration of the examination process. This includes when you leave the room for any reason and also at the end until all students have completed the examination.

- All universities have specific rules about when, or whether, you are allowed to leave the examination room during the examination and you should be informed about these rules prior to commencement of the examination. However, as a general rule within a three-hour examination, students will not be allowed to leave the examination within a specified period usually at the beginning and the end of the examination.

- If you need to leave the room for any reason you should attract the attention of the invigilator and speak quietly informing him/her of your request. If you wish to leave the room for a short time, for example, to go to the loo, blow your nose or because you feel unwell, you will be escorted during this time by an invigilator. If you wish to leave the examination itself, you must advise the invigilator who will take your paper from you and *you will not be allowed to return*.

- If you arrive late to the examination you may be allowed to sit the examination but may not be allowed extra time. If you arrive after other students have left the examination room you will usually not be allowed to enter the room.

5.13 Marking criteria

Examination marking criteria are different to that for essays although the principles underpinning academic knowledge and understanding of the topic remain the same. Examination responses are marked for their breadth of knowledge rather than their detailed discussions. You, therefore, will not be marked down because you do not provide word for word quotations. You should, however, develop your arguments and provide academic evidence to back up the points you make. Ensure that your essays don't simply provide evidence based upon common sense assumptions.

5.14 Feedback

Feedback from examinations is important and helps you to identify your strengths and weaknesses. You are not usually given back examination scripts with comments so if you feel you could have improved in examinations you need to speak to your tutors. They should be able to provide general feedback on how the examinations were graded and identify strengths and weaknesses in your performance.

5.15 Summary

- This chapter has shown why examinations are useful tools for students in that they can test their research and study skills and show their knowledge, and for lecturers so that they can assess whether students have understood course content, and for employers and professional bodies as examination success shows that you have met academic standards required in Higher Education.

- Different kinds of examination were discussed as well as the importance of ascertaining exactly what was expected from each type.

- Planning and preparing for examination purposes in terms of organising revision and gathering and summarising information was highlighted as well as the utilisation of memory strategies and other revision aids.

- Strategies for the examination itself were discussed, with particular reference to writing style, the use of references and more especially the protocols or behaviours that need to be observed during the examination process.

- Marking criteria for examination responses and the relevance of feedback for future work were discussed.

KEY TIP

Write legibly using a black pen in the examination. Pale pink or gold glitter pens may be fine for birthday cards and letters to friends but not for examiners' eyes!

Part Two

Study Skills for Criminology

In Part 2 we begin to examine the skills that will support your studies and link these to the discipline of Criminology. The aim here is to develop an understanding of what Criminology is and introduce some of the debates and theoretical perspectives surrounding the subject matter. This section of the book also looks at some of the research and data collection skills you need to develop if you are to be successful in your studies. Criminological theory is an important aspect of any degree programme looking at crime and disorder, criminal justice or Criminology generally. Chapter 9 focuses on criminological theory and it is the intention that you will begin to develop your skills in applying theoretical perspectives to practical situations. Success in this area of study gives depth to the quality of the work you produce and helps you to acquire analytical skills. These are important for study in Higher Education and also are seen by potential employers as desirable.

The areas covered in this section are:

- **Studying Criminology at University**

- **Finding Information about Criminology**

- **Essay Writing**

- **Studying Criminological Theory and Criminal Justice Policy and Practice.**

Studying Criminology at University

CHAPTER OVERVIEW

By the end of this chapter you should be familiar with:

➤ Criminology as an academic discipline
➤ the other academic disciplines that impact on Criminology
➤ the way in which Criminology is taught within Higher Education.

Criminology is one of the social sciences and is closely related to Sociology, Social Policy and Politics, although increasingly there are links with Psychology and Law. In the main the discipline of Crime and Criminal Justice remains broadly sociologically based. For this reason it can be argued that the key skills necessary for study at this level are transferable; nevertheless, it is important to remember that there are substantial elements of those skills that are used in university that are specific to the study of Criminology. To begin with, it is important that you understand what the discipline of Criminology focuses on and what your motivation is for studying this particular subject.

6.1   What is Criminology?

Criminology is the academic study of crime and disorder, criminal justice systems and responses to crime. It is not necessarily focused on the type of crime portrayed in television programmes and novels where the focus is on investigating crimes and identifying a culprit. It involves a much wider understanding of the various factors that are seen to increase the likelihood of individuals or groups acting in a way that is defined as criminal.

The study of Criminology has been linked with a wide range of other disciplines discussed elsewhere in this book. This means that for some writers, (Tierney, 1996) there is some difficulty in identifying a homogeneous theoretical basis for the study of Criminology. To some extent this is a consequence of the way in which Criminology attaches itself to other disciplines in a parasitic way (Cohen, 1988). The purpose of Criminology can be seen as the way in which common sense notions about crime and disorder are challenged and a broader understanding of these issues developed. While Classical Criminology focused upon the crime committed and the idea of personal autonomy and choice in relation to crime causation, this was later challenged by taking factors other than ideas around human nature and **hedonism** into account. One challenge to this classical thinking was the idea that there were biological and then psychological explanations for crime and the individual criminals. Ideas emerged that involved scientific analysis of human physiological features and physical attributes in relation to crime causation. The emergence of such ideas may seem incredible to us now; however, it is important to remember that genetic research and the search for 'crime genes' continues. Both biological and psychological approaches to understanding crime were highly influential within the tradition of Criminology for a long time. Significantly, these biological factors were prominent in early discussions of crime and, as we discuss in Chapter 9, were based on physical factors.

The development of theoretical positions that look at the impact of social institutions and social interaction in creating our understanding of crime did, however, follow this and these theories have gained more and more credibility as very significant factors in explaining crime and crime causation. A body of writers began to develop theories that explored inequality, social factors and structural factors as the basis for criminal behaviour.

The rejection of individualism led to a series of social explanations of crime as discussed in the work of the Chicago School and Strain Theory (see Chapter 9). This was followed by those theories that emphasise the importance of social control in relation to crime and disorder. Build into this the debates about gender, **sub-cultures** and de-industrialisation and the concept of a free-standing Criminology begins to look weak. The key here is that the study of crime and disorder, deviance and anti-social behaviour cannot be examined in isolation from other disciplines; Politics, Philosophy, Psychology, Sociology and Social Policy are all important areas of study. Throughout this book we have drawn your attention to the range of disciplines that impact on the study of Criminology and it is important that these inter-relationships are recognised.

Criminology as a degree subject involves you in the study of why certain behaviours and activities are deemed to be illegal in some situations and not in others. Criminologists would seek to debate whether there is such a thing as an absolute crime as definitions of criminal activity frequently vary over time and place. It is also an attempt to understand why these actions are deemed to be criminal in the eyes of policy makers and why some groups in society are afraid of the actions of others and as a consequence may feel at greater risk of becoming victims. One example of this relates to the fear of crime; this can be explained as the follows: many older people may feel exposed to crime at a much greater level than statistics suggest – something that is often compounded by the presence of younger people in their neighbourhood (Zender in Maguire et al., 2002). Media representations of the young as disorderly, anti-social or criminal can have an impact on the views of other groups, and can have an influence on their understanding of what these groups of young people are actually like (see Muncie, 2004). Criminologists would seek to examine these ideas about the fear of crime and correlate this with the reality of crime and then seek to provide some understanding of the issues involved.

Some comments that are frequently expressed suggest that solutions to rising levels of recorded crime are nothing more than common sense. Criminologists would counter this by suggesting that a range of factors impacts on individuals and groups and leads to them behaving criminally, and these factors may also have an influence on the way that certain groups are labelled as criminal, anti-social or disorderly. In this sense then Criminology is not unlike Sociology in that there is a need to move away from common sense notions and to develop a 'sociological imagination' (Mills, 1970). In Box 6a we give a definition of criminological theory and in Chapter 9 we introduce you to some theoretical perspectives and ask you to relate these to practical issues surrounding crime and disorder. This is an opportunity to become more analytical and to develop an understanding of the ways in which criminality has been explained by criminologists.

Box 6a Criminological theory

In attempting to explain the facts we know about crime we use theoretical perspectives. Theories are part of that explanation and, according to Vold et al. 'an explanation is a sensible way of relating some particular phenomenon to the ... information, beliefs and attitudes that make up the intellectual atmosphere of a people at a particular time or place' (2002: 2). We discuss a number of theoretical perspectives in Chapter 9.

Another key component of the study of crime is our understanding of criminal justice processes. Box 6b outlines the stages in the criminal justice system. This is an area that can change rapidly but most Criminology courses will study issues like policing, the Court system, imprisonment, punishment and victimology.

Box 6b The criminal justice system

This is made up of a number of agencies who implement the state's responses to behaviour that is deemed unacceptable. The various stages of this process are:

- Policing — this includes investigation, evidence gathering, arrest and charging of suspects.
- The Courts — this involves the prosecution and defence of suspects in a trial and if found guilty, the sentencing of offenders. Also following conviction and sentencing, Courts deal with appeals.
- The Probation Service — this is involved in preparing reports before sentencing, working with offenders in prison and in the community, and supervising offenders who are subject to **community sentences.**
- Prisons — these hold those sentenced to a custodial sentence in a secure environment, work with prisoners to address offending behaviour and to prepare prisoners for their eventual release.
- Increasingly within this process a range of other agencies are involved and the criminal justice system is subject to constant change as policies and theoretical perspectives influence the views of legislators. One example, is that the Probation Service and the Prison Service have recently merged to become the National Offender Management Service (NOMS) reflecting a materialist move within criminal justice agencies typified by measuring risk and controlling populations (see Hudson, 2003).

The Sage Dictionary of Criminology (McLaughlin and Muncie, 2001b) provides further definitions of other topics of study typically examined in this area.

It is important when deciding on the degree you intend to study that you choose a programme that you are going to enjoy. It is also important that the degree is going to cover what you think it will cover. There has recently been a rise in degree programmes that seem to be about Criminology from a vocational perspective but they are often based around entirely different disciplines; Crime Scene Science and Forensic Investigation are examples. Both have a much greater reliance on scientific backgrounds and the investigation of crime rather than an interest in understanding crime in the context of society and social factors. The scientific base of these programmes means that they do not necessarily cover any of the issues typically addressed by a Criminology degree programme studied within a social science department, and it is worth noting that this may also apply to those programmes delivered in Law departments as they focus more on criminal justice and criminal law. These are therefore important considerations when you decide which university offers the programme that most closely meets your interest and career aspirations.

Increasingly, students are able to study a combination of subjects to gain a degree, and so, instead of a single subject programme, their focus may be in two or more disciplines. In most universities the combinations involve two subjects either as major/minor options (see Box 6c), or with each being studied for an equal amount of time as a joint honours degree.

Box 6c What is meant by a major/minor degree?

Major/minor degrees allow you to study two subject areas in combination. The choice of combinations within universities is extensive; however, here are a few examples:

- Criminology with Law
- Criminology with Psychology
- Criminology with Media Studies.

It is usually the case that your programme will involve studying the major option for around two-thirds of the time with the remaining third being devoted to your minor option. The range of choices will vary between universities so if you are looking at this as an option you need to research carefully what is available.

Indeed, in a number of institutions some degrees are only offered in combinations without the opportunity to study a single subject area. This may be in the form of a major/minor route as described above or as joint honours degrees. While it is often possible for students to engage in a wide range of combinations there are some that are more common than others. Sociology, Psychology, Social Policy and Law are quite frequent combinations with Criminology, and recently Media Studies and Forensic Science have increased in popularity. As the responses to criminal behaviour become more inter-disciplinary so the range of degrees has become more varied; however, it is not possible to identify the full range of courses here as the choice is so vast, but some care is needed when choosing the right degree. Another important point is that some degrees focus on Criminology while others focus on criminal justice. This may seem to be a confusing distinction to make. In order to simplify it, examples of the content of several degree programmes are given here. This is the information you are likely to find out about at open days and in the prospectus of the university to which you are applying.

On the UCAS website (http://www.ucas.com), there are four categories for degrees in Criminology:

- **Criminology as a single subject**

- **Applied Criminal Justice Studies**

- **Criminology Law**

- **Critical Criminology.**

In Criminology programmes that are a single subject you may find that examples of the modules you might study are:

- **Criminological Theory**

- **Criminal Justice**

- **Punishment and Social Control**

- **Domestic Violence**

- **Race, Crime and Justice.**

In Applied Criminology programmes examples of the modules could be:

- Social and Public Policy

- Sociology

- Psychology

- Law and Forensic Science

- Criminal Justice.

In Critical Criminology programmes combined with Law you might study:

- The Critical Analysis, applied to Criminal Justice

- Legal Methods

- Law of Contract

- Criminal Law.

A programme combining Criminology and Law might include:

- Introduction to Criminology

- Drug and Alcohol Awareness

- English Legal System

- Introduction to Crime and the Criminal Process.

These are only examples drawn from the range of material presented by universities to describe their programmes and you will need to make your own enquiries about the content at the university of your choice. They do, however, provide you with an indication of the range of modules that are taught on Criminology degree programmes.

The range of study skills that you will need to study any of these options is discussed in this book and applies to all of these areas as well as to the variety of combinations that are available.

6.2   How does Criminology relate to other disciplines?

As mentioned above Criminology is regularly studied as part of a combination with other disciplines; however, even when Criminology is studied as a single subject there is considerable overlap with other disciplines. Criminology programmes are often based upon either a sociological, psychological or socio-legal perspective; however, these programmes also usually draw upon a number of social science subjects because, as noted by Cohen, 'somewhat like a parasite, criminology attaches itself to its host subjects' (1988: 4). These host subjects include: Sociology, Social Policy, Politics, Law and Psychology.

An understanding of sociological theories and debates about how societies function and control populations is clearly crucial to understanding crime and criminal justice. Giddens defines sociology as 'the study of human social life, groups and societies. It is a dazzling and compelling enterprise, as its subject matter is our behaviour as social beings. The scope of sociological study is extremely wide ranging, from the analysis of passing encounters between individuals in the street to the investigation of global social processes' (1997: 2). Therefore when thinking about Criminology we think about how, if at all, crime is shaped by societies – is crime located within the individual or within societies? Or we can go a stage further by exploring how societies shape an individual's thinking (see Durkheim on suicide). This is crucially related to how we tackle crime – do we address individuals, for example, through punishment or do we address society, for example, through tackling poverty or do we need to address both? (This is discussed further in the extract in Activity 9g). Another example is the ways in which globalisation affects crime – think, for example, about drug importation and how, if at all, this can be controlled in an increasingly global market.

Social Policy is another subject area that overlaps with criminal justice in particular, but also with some areas of criminological theory. Some writers (Garland, 1985; 1990; 2001) have suggested strong links between welfare (social policy) and crime (or social) control. Political ideology has also influenced our understanding of social development and this is especially relevant in the context of why and how we punish offenders.

Law is another area where the overlap is fairly evident, in the context of criminal law at least; however, there are also implicit links with other forms of legislation. A good example of this is whether changes to the legislation relating to licensing hours will affect alcohol-related street violence and disturbances. We can also think about how legislation changes over time, and between cultures and societies reflecting the 'norms' and 'values' at a given point in history. As criminologists we can question and debate law and explore the variety of competing explanations that underpins the rationale for legislation. Furthermore, we can explore the theoretical backdrop for legislative change over time. For example, the welfare versus justice models of criminal justice.

The above debates about law and legislation relate to wider debates about politics and philosophy. A clear example of this relates to drugs legislation and state control. Is it right that the state intervenes in an individual's decision to partake in risk-taking behaviour? Moreover, definitions of risk are subjective and are open to numerous factors, such as risk to the individual, to communities, to society, to the planet, financial risks and many more. Therefore, creating a hierarchy of risk-taking behaviour is very difficult.

Activity 6a

Name ten legal and illegal drugs and place them in order of harmfulness starting with the most harmful. You also need to define what your criterion is for harm.

There are no right and wrong answers for this activity, but it will help you think about the influences of social factors, politics and cultural 'norms' and 'values' on legislation and current thinking regarding what is acceptable and unacceptable behaviour in contemporary society.

You may also find it helpful and interesting to complete this activity with a friend or friends.

It is likely that any study of Criminology will include elements of these and other disciplines; indeed, as the subject area develops, other disciplines may be appropriately linked to Criminology to further your understanding of key debates. The important message here is that crime is not something that occurs in a vacuum, without any relationship to other events, but that crime is constructed as a consequence of a range of social, psychological or environmental factors. These factors are the basis of the theoretical perspectives that are discussed in Chapter 9.

KEY TIP

Remember to make connections between the various disciplines as this gives greater depth to your understanding of Criminology.

6.3   Summary

Hopefully, having read the preceding section you will still remain interested in studying Criminology at university. You will appreciate that crime is something that is connected by a number of factors to other aspects of social life and academic study and that the solutions to the perceived problem of rising crime is more complex than a simple debate between notions of punishment or treatment.

Some of the topics highlighted here have focused on:

- **what do you study when taking your degree in Criminology**

- **whether this is done independently or relates to other subject areas**

- **the importance of making connections with other topic areas throughout you degree.**

Finding Information About Criminology

CHAPTER OVERVIEW

By the end of this chapter you should be familiar with:

➤ reading for a purpose
➤ different sources of information
➤ finding information.

The aim of this chapter is to focus upon sourcing and using information relevant to criminological study. In recent years with the increased use of Information Technology, the breadth and range of information that is quickly available to students and academics has risen dramatically from electronic journals, and electronic databases to websites and publications on-line.

The chapter will initially focus upon why you should collect information before going on to explore ways of reading, storing and using information. The chapter will illustrate why it is important to start reading for study with an organised system in place. Following this, the chapter will briefly show you how to find books and journals in your library. Finally, we will provide you with a key skill check and further reading.

7.1 Why should you collect information?

It may seem a ridiculous question, yet it is vital that we consider this before we start, as this will affect how we go about collecting information. The first, and most obvious answer, is to learn about something new – thus we read for interest. If this is your purpose then enjoy your reading.

Reading at university, however, is often assessment related and therefore we need to think about what we are trying to do and find out. Within the vast majority of assessed criminological work you will be trying to argue a point and this involves critical analysis of ideas, arguments and writings. If we think about an argument, a successful one involves the production of evidence that will support a particular viewpoint rather than your own personal opinion, which is often not supported by reasoned analysis. Within your academic work then you need to avoid opinion and instead base your arguments on the analysis of evidence.

Reading to find evidence for assessed work is, therefore, very different from reading for background information or pleasure. Reading for evidence is a process and this chapter will now go on to examine this process and suggest methods that you may find useful.

7.2 Starting out

You will frequently start out with a question and the initial reaction is often to go straight to the library and search for books with titles that contain the chosen topic or hunt for the relevant books from your reading lists. From these, you make reams and reams of notes and then write the essay. With this method, it is not uncommon while writing the essay that you sit there thinking, 'I'm sure I've read something on that somewhere', and then the search through the notes begins. This is a very ineffective and time-consuming process and often leads to the production of disjointed work that jumps from one topic to another without any clarity or plan.

A more effective way, however, is to embark upon your information retrieval with a strategy of what you are aiming to do. Often the best way to start is with a textbook chapter on your chosen topic. As you read the chapter, jot down the main debates. From these ideas you can start to think about the key issues you want to cover within your assessed work.

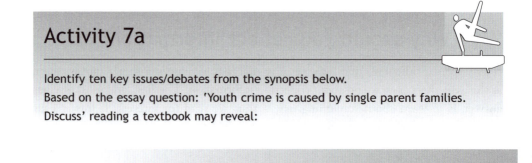

Activity 7a

Identify ten key issues/debates from the synopsis below.
Based on the essay question: 'Youth crime is caused by single parent families.
Discuss' reading a textbook may reveal:

- Crime causation is complex. Within such debates 'crime' is used in a narrow sense, for example, the focus is upon visible crimes and anti-social behaviour.
- Single parent families includes a multitude of family types, for example, widows, divorcees, those who have never married. This complicates debates about single parent families and relationships to crime.
- Some argue families headed up by single mothers are more likely to have criminal children because of a lack of discipline within the home.
- Some argue that this lack of discipline and single parent families is linked to moral decline and the rise of an **'underclass'**. Some argue that it has less to do with single parenting and more to do with where some single parents find themselves living, i.e. poor housing in areas with high levels of unemployment, truancy, crime and drug use.

(see Appendix 1 for a sample answer to this exercise.)

Once you have identified the key debates then it is helpful if you start to organise the information. One useful way to do this is to use a lever arch file and some file dividers. Label the dividers according to the key debates identified from your initial reading. Once you have done this, you are beginning to structure your thoughts and your working patterns. Then as you read, write your notes into the appropriate sections. You may also want to write out quotations into the appropriate sections too. You may want to be more flexible with your reading style; you don't have to read every paragraph and every word. Skim read the text, pull out and focus upon the material that is relevant to what you are doing.

KEY TIP

Remember to fully reference the quotes even at this stage (see section 8.6). This will make things easier in the long run and ensure that you do not plagiarise the work of others. Writing page numbers within the margins of your notes with full bibliographic details is a good reading habit to get into.

7.3   Reading for a purpose

Now you have identified the key debates and started organising your work, you can start to read for a purpose; thus, rather than simply looking for books with the key-words in the title, you are starting to look for more specific forms of information as your thoughts and ideas evolve. For example, although the question in Activity 7a is focusing upon single parents and youth crime, you may find yourself reading about the relationship between unemployment and crime as it may be the case that living in an area with high unemployment is equally or more important than whether a young person is from a single parent family. Alternatively, it may be that being unemployed and being a single parent are both contributory factors or that the relationship between the two is too complex to discuss in terms of causality.

One way to progress your reading is to follow up the texts cited in the textbook chapter. If a text is referenced numerous times in a chapter then you can usually assume that this is a key text. Go to the bibliography and find the full reference details for the text. Another way to move forward is to look for texts with the new keywords in the title or search the journal databases for appropriate articles (how to do this will be covered later in this chapter).

As your reading progresses you are starting to collect a range of evidence and ideas connected to the topic. Moreover, your reading is no longer general; instead, you are reading for a purpose – to find out particular bits of information to record in your file. The structure of your work is shaping up and this will be of great help when you come to write up the assessment.

There is no magic number in relation to the number of books and/or articles you should read. But remember that this is *your* work and *your* argument that you are trying to present and, therefore, relying on one or two books will inevitably lead to the presentation or summarisation of others' work. You will get a feel for when you have read enough because you will start to see repetition; more often than not, however, the amount of reading you do, will be based upon pragmatic reasons relating to time. Using the file system will allow you to identify where you need to do more reading and research. If, for example, you have very few notes in one area that you have identified as a key topic then you know that you need to focus your reading in that area.

The above is only a suggestion of how you may go about reading, organising your note-taking and identifying further reading. While you may not adopt this approach the important thing to remember is that you should try to organise your reading and note-taking from the outset and that you need to develop your own system for doing this. Always try to use a broad range of literature and do not just stick to texts that contain your keywords in the title. If you are studying a range of modules think about how the modules overlap and try to make links with your reading and note-taking.

KEY TIPS

Always start your reading early in the course. The supply of library books will be limited and if you leave your reading until just before the assessment has to be completed you will find all the books are on loan.

If you are struggling on your literature search, one useful strategy is to use the bibliography of a useful chapter or article to search for further resources.

7.4   Sources of information

There are numerous sources of reading information that you may draw upon in your collation of material while studying Criminology. Within academic reading we must always be concerned with *validity*. We must, therefore, constantly ask the question: how reliable is the source of information? Earlier in this chapter, the importance of using evidence to back up points or arguments was highlighted. If, for example, a researcher conducted a **longitudinal study** that examined official police arrest rates of heroin users and this was then used as the sole piece of evidence to illustrate a rise in heroin use in the particular locality in question, one may critique this on the grounds of validity. While the evidence may show an increase in arrest rates, this does not necessarily mean that heroin use has risen; it may, instead, be the case that the police have targeted heroin users thus leading to an increased arrest rate. Often such figures tell us more about society's reaction to crime rather than the crime itself.

In this example we have used the term longitudinal research and it is worth explaining what we mean by this term. Longitudinal research is where a research study takes place over a period of time. The researchers are often interested in change over time so they interview people, for example, several times over several years. Examples could be following all children born in a particular year through various stages of their life course.

This section of the chapter will now go on both to look at the various sources of information that you can use in your study of Criminology, and also to consider the validity of these sources.

TEXTBOOKS

Textbooks are a very good place to start and can provide you with an overview of the chosen area. They are academically valid and will be of great help to you during the course of your studies. One thing that you have to bear in mind, however, is that they are secondary sources and that they are the author's interpretation of the work of others. (This is also discussed in Chapter 8). Depending upon the level and depth of your study, you may want to go to the original sources to see if (a) your interpretation is the same and (b) to find out more detail about the original ideas or research.

You will usually find textbooks on the reading lists associated with particular modules and purchasing some of them is often a useful investment as you can draw upon them across a range of modules throughout your time at university. As textbooks are written as aids to learning, you will find that there are several levels of sophistication and complexity. For example, lecturers on Level 3 modules do not usually recommend texts that are recommended to you in Level 1. Moreover, you may find that some texts are too complex when you start your studies – don't let this put you off. More often than not students who feel like this have selected texts that were written with Level 3 or postgraduate students in mind. Ask your tutors for advice before you rush out to the bookshop.

Examples of texts suitable for Level 1 include:

Cavadino, P. and Dignan, J. (2002) *The Penal System*, 3rd edn. London: Sage.

Downes, D. and Rock, P. (2003) *Understanding Deviance*, 4th edn. Oxford: Oxford University Press.

McLaughlin, E. and Muncie, J. (eds) (2001) *Controlling Crime*, 2nd edn. London: Sage.

Muncie, J. (2004) *Youth and Crime*, 2nd edn. London: Sage.

Muncie, J. and McLaughlin, E. (eds) (2001) *The Problem of Crime*, 2nd edn. London: Sage.

Tierney, J. (1996) *Criminology: Theory and Context*. London: Prentice Hall.

Williams, K. (2004) *Textbook on Criminology*, 5th edn. Oxford: Oxford University Press.

RESEARCH MONOGRAPHS

These are texts that are based upon research by the author(s). They are very good detailed sources of information; however, unlike textbooks, they are often very narrow in their focus. You may want to cross-reference these with other sources of information as they do not always provide you with a breadth of knowledge from across the topic that you are studying.

What research mongraphs do provide you with is a detailed insight into particular areas. Because the texts often deal with a defined topic they explore the issues in much more depth. The texts provide arguments based upon research and you can decide whether you agree or disagree with the arguments and the research evidence provided.

Examples include:

Becker, H.S. (1963) *Outsiders: Studies in the Sociology of Deviance*. New York: Free Press.

Parker, H. (1974) *View From the Boys: A Sociology of Down-town Adolescents*. Newton Abbot: David Charles.

Pearson, G. (1983) *Hooligan*. London: MacMillan.

Winlow, S. (2001) *Badfellas: Crime, Tradition and New Masculinities*. Oxford: Berg.

JOURNAL ARTICLES

Similar to research monographs journal articles are based upon research by the authors. This research can either be library based, and, therefore, the article is based upon ideas, theory and philosophy, or they can be based upon empirical research (fieldwork). These are excellent sources of information and like research monographs are often very specific. Try to use articles taken from peer-reviewed journals. Peer-reviewed journals are journals where the articles are submitted to the journal and reviewed by a panel of academics before they are accepted for publication – this ensures that the publications are of a high quality.

Examples of key journals include:

British Journal of Criminology

Howard Journal of Criminal Justice

Theoretical Criminology

Youth Justice.

NEWSPAPERS/NEWSPAPERS ON-LINE

Newspapers can be a useful source of information within Criminology and can be used for a number of different reasons. They should, however, always be used with caution. As pointed out by Muncie:

> Despite the powerful 'commonsense' view that news media merely provides the *facts* of a process in which crime occurs, police apprehend criminals and courts punish them, the relationship between crime and media reportage is far from simple. (1999: 44, original emphasis)

Newspapers should therefore be used very carefully. All newspapers have a political bias and this must be taken into account when taking on board the information. Furthermore, as MacDougall (in Hall et al., 1978: 52) points out:

> At any given moment billions of simultaneous events occur throughout the world ... All of these occurrences are potentially news. They do not become so until some purveyor of news gives an account of them. The news, in other words, is the account of the event, not something intrinsic in the event itself.

Newspapers often present atypical stories in a manner that makes them seem like common everyday occurrences. Just this morning, for example, I read that 'carjacking is now very common in the UK'. While there may have been an increase in this type of crime it can hardly be described as 'very common'. Chibnall (1977: 77) argues that there are five rules that shape whether a story will get published:

1 visible and spectacular acts

2 sexual or political connotations

3 graphic presentation

4 individual pathology

5 deterrence or repression.

He goes on to argue that we must always be aware of two things: firstly, that a selection process has occurred and, secondly, that a certain presentation style has been adopted. Therefore, when using newspaper reports within academic work we must always bear these things in mind.

Newspapers can be broken down into two main groups, tabloids and broadsheets. Both can be useful resources, but for different reasons.

Broadsheets

These include papers such as the *Guardian*, *The Times*, *The Independent* and the *Telegraph*. These will help you understand the importance of politics to Criminology and will keep you up to date with new crime and criminal justice legislation, policy and practice.

Tabloids

These include papers such as the *Sun*, the *Sport*, the *Mirror* and the *News of the World*. These papers can be useful for media analysis but should not be used for information relating to the size and nature of crime in the UK and the rest of the world because the stories are often exaggerated and misleading.

Information Technology has transformed the usefulness of this source material. In the past, students were reliant upon paper copies or microfiche to source information and this was a very time-consuming method. Since then newspapers have been stored upon CD-ROMs which are much more accessible or in very recent years a number of newspapers have set up their own Internet sites. These sites are very good, particularly in relation to searching for old articles. However, more often than not, accompanying pictures are not included with the archived text. This is unfortunate because, for example, with the study of crime and the media, we are often interested in the graphical representations used as this provides us with an insight into how the reports have focused attention on particular aspects of the story. For example, pictures might focus on those aspects of an event that are the most emotive while overlooking some of the evidence that alters our views towards the offender.

Activity 7b

Compare and contrast four different newspaper accounts of the same crime story. What does this tell us about the validity of newspaper accounts?

WEBSITES

There are a whole range of websites related to Criminology, some exceedingly useful, such as the Home Office site (www.homeoffice.gov.uk); others are not so useful. There has certainly been a growing tendency for students to use electronic sources of information over recent years and while this has great benefits there are also many negative aspects to this trend. The most important thing we need to think about when using a website is that of academic validity. While some crime websites may appear interesting and even seem to contain some useful facts about crime, the validity of such facts may be questionable. Thus, you need to ascertain if the content is valid – Where is the material from? What is the evidence to back up such statements? In light of this, if you have found a website and you are not sure of its status check with your tutor.

> **KEY TIP**
>
> When discussing crime and criminality within a UK context avoid using information from dot.com sites that relate to the US. Always ensure that the information you are using is relevant to the context you are discussing.

WEB PUBLICATIONS

As well as websites, you will also find a number of publications on-line. Many of these publications are in a pdf format and you can download programmes, such as Adobe Acrobat Reader, free of charge. These programmes will allow you to download the reports and read them. There are numerous academic research reports available through the Home Office website.

The various Home Office websites are also very good for downloading or viewing government reports or Acts of Parliament, for example, the Crime and Disorder Act 1998 can be found on the Home Office website. These publications are excellent resources for the criminal justice element of your studying.

USEFUL WEBSITES

The following is a sample of a few fairly general websites relating to Criminology that you should find useful. You will find more specific websites relating to individual topics, such as domestic violence, prisons and drugs, in Appendix 6.

http://www.britsoccrim.org

This is the website for the British Society of Criminology (BSC). You can become a member of the Society if you engage in postgraduate study. The site is useful for everyone because it contains information about conferences, and it includes the BSC's ethical guidelines as well as selected papers from a number of annual conferences.

http://www.kcl.ac.uk/depsta/rel/ccjs//home.htm

This is the home of the Centre for Crime and Justice Studies (CCJS) which 'is a charity which aims to inform and educate about all aspects of crime and the criminal justice system from an objective standpoint, and in accordance with the Centre's values' (www.kcl.ac.uk/depsta/rel/ccjs). You can become a member of CCJS and as a member you receive the quarterly magazine *Criminal Justice Matters*. At the time of writing UK student membership is £20.

www.homeoffice.gov.uk

This is a very useful website which contains voluminous quantities of information. The site is broken down into numerous sections which are almost like websites within websites. The website is currently broken down into ten areas and you may find the following particularly useful:

- **Community and Race**

- **Crime and Policing**

- **Justice and Victims**

- **Drugs**

- **Terrorism**

The Home Office funds a great deal of criminological research and most of the findings are published as pdf files on this site. While no one would doubt the quality of the research that you find on this site you must always consider who has funded the research and how this may affect the research focus.

http://www.poverty.org.uk

'This site monitors what is happening to poverty and social exclusion in the UK and complements our annual monitoring reports. The material is organised around 50 statistical indicators covering all aspects of the subject, from income and work to health and education' (www.poverty.org.uk).

http://www.sosig.ac.uk

This is the site for the Social Science Information Gateway (SOSIG). The site 'aims to provide a trusted source of selected, high quality Internet information for researchers and practitioners in the social sciences, business and law. It is part of the UK Resource Discovery Network' (sosig.ac.uk).

www.jrf.org.uk

This is the website for the Joseph Rowntree Foundation (JRF). 'The Joseph Rowntree Foundation is one of the largest social policy research and development charities in the UK. It spends about £7 million a year on a research and development programme that seeks to better understand the causes of social difficulties and explore ways of better overcoming them. The Foundation does not carry out the research in-house, but works in partnership with a large variety of academic and other institutions to achieve its aims' (www.jrf.org.uk). This site includes a useful section on JRF press releases, many of which concern findings from JRF research projects. The site also contains a very good web links section.

http://www.socialexclusion.gov.uk

This is the site of the government's Social Exclusion Unit. The site contains useful information on social exclusion including news and research publications.

http://www.ukcjweblog.org.uk

This useful website provides 'the latest news about criminal justice issues from around the UK, drawn from media websites, government sources and criminal justice organisations' (www.ukcjweblog.org.uk)

http://www.hero.ac.uk

This is the site for the Higher Education & Research Opportunities in the United Kingdom (or HERO for short!). According to the site: 'Whatever your interest in higher education or research in the UK, we're here to help you find out more. HERO contains information on all aspects of higher education across six sections packed with information, advice and essential contacts' (www.hero.ac.uk). The site contains loads of information from choosing a course and studying at university to finding a job.

Newspapers on-line

You will also find various newspapers online and these include:

http://www.guardian.co.uk	The *Guardian/Observer*
http://www.timesonline.co.uk	*The Times*
http://www.independent.co.uk	*The Independent*
http://www.dailymail.co.uk	The *Daily Mail*
http://www.express.co.uk	The *Daily* and *Sunday Express*
http://www.thesun.co.uk	The *Sun*
http://www.dailystar.co.uk	The *Daily Star*

Electronic media websites

There are also a number of electronic media websites where you can keep up to date with the news and these include:

http://news.bbc.co.uk	BBC News
http://www.channel4.com/news	Channel Four News
http://www.itn.co.uk	ITN News
http://www.sky.com/skynews/home	Sky News
http://www.reuters.co.uk	Reuters UK News

7.5 Using a variety of sources

Within your work it is always beneficial if you can use a variety of the sources outlined above. If, for example, you are discussing the use of imprisonment and you are drawing upon Cavadino and Dignan's text *The Penal System*, try to interlink the arguments with supporting material from articles in journals such as *The British Journal of Criminology* or the *Howard Journal of Criminal Justice*. If you discuss the size of the prison population, evidence this with figures from www.hmprisons.gov.uk, for example.

Your arguments will be better evidenced if you can draw upon a range of sources and use these within your work. Furthermore, using a range of sources such as journal articles and Internet publications can mean that you are drawing upon contemporary sources. This is really important when discussing aspects of the criminal justice system, for example, because new legislation comes into force all the time. If the text you are drawing upon is outdated, although some of the theoretical arguments are sound, the evidence is out of date and this can make your work appear out of touch. It annoys assessors when students refer to Home Secretaries who left the post five years ago, particularly if they lost their post because their party was not re-elected (thus, not only is the Home Secretary wrong but so too is the political party holding office).

7.6   Finding information

So far this chapter has focused upon methods of reading, note-taking and different sources of information. The chapter will now go on to explore how you go about finding this information, focusing upon books and journals.

FINDING A BOOK

When you start studying you will be given an indicative reading list, usually located in a module or course guide. This reading list is a guide and therefore you do not have to read every book on the list, but likewise it is an indicative guide and therefore should not be taken as a list of the only books you should read.

When you start university you will be given a library induction and within this you will be shown how to locate books. Higher Education libraries work using a catalogue system to order, store and shelve books. While the exact nature of this system varies from one establishment to another, the basic principles are the same. You interface with the catalogue via an electronic database (OPAC: Online Public Access Catalogue) where you can enter the author's name, the book title or a keyword. The system will then present you with a list of matching texts with reference numbers relating to the shelving system (usually referred to as the shelf mark number). You simply make a note of the reference number and locate the book and, hey presto, you can begin reading!

If the book is on loan it is usually possible to reserve the book so that when it is returned you can take the book out. Remember though, the reservation system is only as reliable as the person who has the book (many university libraries operate fines to 'encourage' people to return their books on time). If they fail to return the book on time, then this has a knock-on effect on you. Thus, to reiterate a point made earlier, it is vital that you allow time for your reading.

KEY TIP

When searching using the 'keyword' feature use a variety of terms for the topic you are wanting to search for: for example, illegal drugs, illicit drugs, illegal substances, illicit substances, narcotics.

FINDING A JOURNAL ARTICLE

If your reading list contains a journal article, you locate it in the same way as a book. Use the OPAC and enter the title of the journal, retrieve the reference number and then physically locate the journals and find the edition you want.

If you do not have a particular article in mind and instead you want to conduct a key-word search, the process is a little different. Finding a journal article is slightly more complex than finding a book. The OPAC system allows you to search for a journal title but not for an article title. Therefore, to find articles on the topic you are interested in, you need to go through a separate database.

There are a number of on-line databases that contain indexes to journal articles, many often include abstracts (summaries) of the articles; some also include the full text of the journal articles themselves. Hyperlinks to these databases are usually contained on the university's library website. You search these databases in the same way as the OPAC using a keyword, author or title. Unlike the OPAC, however, the articles that match your search may not necessarily be in your library (check using the OPAC for this). Box 7a contains a list of useful databases.

Box 7a Examples of some resources available electronically

ASSIA (Applied Social Sciences Index and Abstracts) is a bibliographic database which gives references (with abstracts) to the literature of the applied social sciences, including social policy, social services, society, family issues, economics, politics, employment, race relations, health, education and youth work. The abstracts are taken from over 550 English language journals from 16 countries.

BHI (British Humanities Index) is a bibliographic database which gives references, with brief abstracts, to all areas of humanities. It covers from 1962 onwards. Approximately 400 journals are included and it is updated monthly.

ISI Web of Knowledge — **ISI Web of Science** consists of the three ISI databases (Science Citation Index, Social Sciences Citation Index and Arts & Humanities Citation Index).

PsycINFO contains citations and summaries of journal articles, books, technical reports and dissertations in the field of Psychology and related disciplines, such as Psychiatry, Physiology and Sociology. Over 55,000 references are added annually.

ScienceDirect is the most comprehensive database of primary literature available in the sciences. It contains the full text of more than 1,700 journals in the life, physical, medical, technical, and social sciences. It also contains abstracts from the core journals in the major scientific disciplines.

Zetoc provides access to the British Library's Electronic Table of Contents of around 20,000 current journals and around 16,000 conference proceedings published per year.

This chapter cannot give you explicit instructions on exactly how to conduct these searches because different universities operate different systems. The important thing to remember is that if you cannot do something or you need support and assistance ask a member of staff in the library. If you can operate the systems and find the information you need early in your academic career, these skills will be vital as you will utilise them throughout your time at university. Furthermore, if your retrieval skills are sound then you will find that writing your assessments is easier as your work has made use of a broader base of evidence than simply relying upon one or two textbooks. Assessors like to see students use a broad range of sources and the use of journal articles alongside books will be particularly welcomed.

Once you have found the articles, you can either find the hard copies or many journals are now available on-line. Some of the journal databases allow you to link straight to the journal on-line. If you cannot do this you can link to the journal through the university's website. The on-line journals websites also allow you to browse the journals or search within the journal itself. One of the great things about the on-line journals is that you can access them from home. In order to do this, however, you will need a password or passwords. As noted above, it is useful if you can familiarise yourself with these processes as early in your academic career as possible.

KEY TIP

Ensure that you request in advance a password(s) for the on-line journals you want to use as it sometimes takes a while to process the requests.

Activity 7c

Conduct a journal search and locate five articles in your library relating to race and crime, or choose a topic that you are interested in.

7.7 Summary

- This chapter has focused on why you should read at university and has suggested a model that you may find useful to help you organise your reading and note-taking. While you may decide not to adopt this model the important thing to remember is that you need to have a system which ensures that you are organised in your reading, note-taking and identification of further reading. The first book you read is the start of a process and should not be viewed as the start and finish.

- The chapter has highlighted the range of sources that you can draw upon and has drawn attention not only to the importance of evidence within your arguments but also to the fact that the evidence you use needs to be academically valid. Make sure that you organise your note-taking and fully reference the work even in your notes! You *must* cite the work of others; otherwise your work may be seen as plagiarised.

- Finally, you should enjoy your reading. Remember to find a good place to read (not in front of the TV), and take breaks. If you read the same few lines over and over again or get to the end of a paragraph and think to yourself, 'I haven't a clue what I have just read', then you know it is time for a break.

Activity 7d: Test yours skills

1 Find four books about the criminal justice system and record the shelf mark number.
2 Locate two journal articles about the fear of crime.
3 Write a 500-word review of one of the articles.
4 Compare the facts and figures in two on-line newspaper articles about a crime-related story.

8 Essay Writing

CHAPTER OVERVIEW

By the end of this chapter you should be familiar with:

➤ essay preparation
➤ note-taking
➤ data collection
➤ references and bibliography
➤ essay writing
➤ presentation
➤ marking criteria
➤ dissertation writing.

Whatever type of assessment you are required to complete for particular Criminology modules, all will involve written work. Essay writing is an opportunity for you to develop an extended discussion of a topic area. You will seek to demonstrate your understanding of the question and the academic debates that surround it. This is not simply your own opinion in relation to the subject area but is a way of showing that you can formulate a coherent and informed debate based upon a range of relevant literature. Knowing how to structure and communicate your thoughts in written form is a necessary skill for any student. Being able to write well, however, is not some sort of natural gift that we are born with; it is a skill or technique that can be learned. (Rudestam and Newton, 2001: 204) More than this, your writing style may need to be adapted depending on which subject you are studying and what type of assessment is required.

> **KEY TIP**
>
> Most universities would expect students from within the social sciences to write in the third person rather than use the first, as it de-personalises the discussion and produces a more objective piece of work. So for instance, rather than writing 'I think', 'I suggest', 'I would argue', in the first person, most tutors would expect you to write – 'it will be important to examine', 'this essay will explore', 'consideration will be given to', in the third person. Criminological writing is almost wholly written in the third person.

We have already identified that, in order to gain a university degree, it will be necessary for you to produce written work, and knowledge of planning, structuring and ways of conveying what you want to say will aid effective written communication. Being an adult learner has both strengths and weaknesses in relation to study: for example, one of the strengths that those of you who are mature students may have is that you bring with you lots of experience which will be useful in completing a degree; however, a weakness for mature students returning to learning might be the competing demands upon your time such as balancing work, family and university commitments. Those of you who have come to university straight from school or college might feel that one of your strengths would be that you are continuing and progressing in your studying and that the recent experience of undertaking examinations is useful; however, a weakness might be that you also have to undertake paid employment.

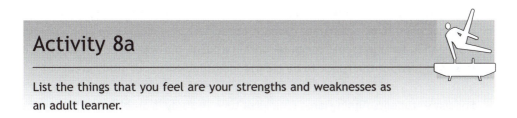

Activity 8a

List the things that you feel are your strengths and weaknesses as an adult learner.

Understanding where your weaknesses are is useful in allowing you both to reflect on those areas on which you need to allow more time, work or effort in order to improve and also to organise your workload and other commitments. Knowing where your strengths lie is equally important in the sense that it will allow you to build upon your confidence and improve even further in these areas.

8.1 Planning your essay preparation

Within the time frame of the particular module you are taking you need to make sure that you allow yourself enough time to read, research, take notes, draft, amend and submit your essay. This will be just one assessment of many and you will need to prioritise your assessments in order to meet the different deadlines for the different tasks. To begin with, consider Activity 8b.

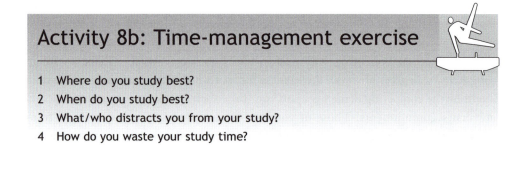

Activity 8b: Time-management exercise

1   Where do you study best?
2   When do you study best?
3   What/who distracts you from your study?
4   How do you waste your study time?

Bearing your answers in mind, consider the following:

- For many students having a quiet and dedicated space to study is important. This does not need to be a separate room. It can in fact be a cleared writing space or workstation for a personal computer within a room, but wherever it is located, it is a good idea for the space to be clearly identifiable as your study area.

- For some students finding large chunks of time to study is problematic if they are juggling many varied commitments, such as family and paid employment as well as study. Being organised in terms of time management is vital for such students (and most lecturers!) as they will need to balance competing commitments. Of course, social time for watching television and meeting with friends should be allotted time spaces but this time may need to be limited. Friends and family can distract from study time. However, our relationships with others are important and therefore some organisation for socialising is useful. Students may also have to rely on family and friends to support them with personal responsibilities such as child-care or care of another relative or friend. Managing a personal life with partners, family, friends, work and study commitments can be difficult and time-management strategies are valuable tools to use to help us keep a sense of balance.

- Many of our study skills undergraduates have explained that rather than getting down to allotted tasks at once they often put off making a start on a piece of work. The reasons for this can vary from it being too daunting a task and not understanding the task completely to it being one task on a long list of many things to be done within a tight time frame. These students are not alone — many of us will put off certain tasks for a variety of reasons. Some will clean ovens rather than start a specific task only to find themselves more stressed the next day even though the oven is clean! In such cases it is important to try to cut large pieces of work down into manageable chunks — so, for instance, if it's a case of writing a Criminology essay, write a checklist of things to do in order to get the essay completed. This can be helpful in the sense that now it is a list of ten small tasks rather than one huge and quite vague task.

- Another significant factor that often prevents the start of essay writing is sometimes referred to as the 'blank page syndrome'. This is when you are unable to find a starting point. When this occurs no amount of time management seems to help. However, there are ways of addressing this. One way is to jot down some keywords to see if you can make some connections between them. It does not matter if ideas do not immediately tumble out onto the paper and readily cohere into an essay; the point is that if there is nothing there in the first place there is nothing to work with so putting comments and notes down on paper is at least a starting point and does often stimulate thoughts and ideas. For more tips about how to get started consider Jody Veroff's chapter in Rudestam and Newton (2001: 204–5).

Your check list may look like this:

- Set out a time frame bearing in mind the deadline for submission. Allow yourself a little flexibility for things to go differently.

- Go to the library to begin the research and find the resources necessary to write the essay.

- Begin reading and taking notes from the resources collected.

- Make a plan of how you intend to answer the essay — this will help to keep you on track and help avoid going off on interesting, though unhelpful, tangents.

- Begin to write a first draft of your essay.

- Edit and write up a full essay answer — type this up checking spelling, grammar, structure and references.

- Type up your bibliography.

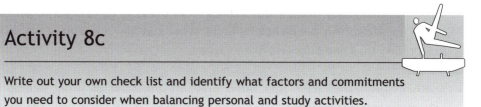

Activity 8c

Write out your own check list and identify what factors and commitments you need to consider when balancing personal and study activities.

Examples may relate to family, work and child-care, or you may have sporting, leisure and recreational commitments.

KEY TIP

Keep a diary for personal and professional tasks and activities.

8.2 Reading and understanding the question

When you have decided which question you are going to answer, read the essay title carefully, making sure you fully understand what the essay title is asking of you. Break the question down into components that show what the requirement is. Deconstructing the essay question in this way is very important. You should pick out the key words and phrases that are central to the question as your answer must focus on these points.

KEY TIP

Choose a question that you find really interesting. You are more likely to be motivated to research it.

Typical examples of questions relating to Criminology could be:

1 'Slowly but surely public confidence in the British police is being eroded.' Discuss.

2 Critically discuss the concept that restorative justice is nothing more than rehabilitation repackaged.

3 Compare and contrast some of the principal justifications for punishment.

4 Critically evaluate the contention that 'prison works'.

While all of these examples would require you to demonstrate an understanding of key issues and to be able to describe them, they require you to go further than this. In the first example, while you are asked to discuss public attitudes to the police, you would also be required to identify a variety of perspectives that surround this topic area. In comparison to this, the second question requires you to develop this even further and give critical discussion; this means you will be assessing strengths and weaknesses of the various perspectives and ideologies that you have identified as being relevant to this topic.

In the third question, compare and contrast means exactly what it says! However, this does not simply mean you would only describe the principal justifications for punishment (for example, **incapacitation**, **deterrence**, retribution, and rehabilitation) but that you would seek to show what similarities there might be and what differentiates them or sets them in opposition to each other.

Critical evaluation is asked for in question four and this would involve interrogating all sides of the argument. In this example, you could be examining the fact that a rise in prison population has coincided with an apparent decrease in criminality. A critical evaluation would seek to either confirm the correlation or suggest alternative explanations for both of these statistical facts.

Critical thinking is about not taking things at face value. Rather than accepting information that is presented as authentic or as 'the way things are', critical thinking demands a more perceptive and rigorous investigation of the perceived 'fact'.

8.3   Gathering information for your essay

KEY TIPS

Make sure you know the required word length for your essay and stick to it – you may lose marks if you do not meet the target set.

Get into the habit of using your word count key in the tools section on your toolbar to quickly check the number of words used in your work.

Having chosen the question you wish to answer, the next step is to gather information about the topic. The previous chapter has already introduced you to reading in Higher Education and ways of accessing relevant information. This section will focus upon specific reading techniques necessary to gather the relevant information and knowledge you need to produce academic work. In this context reading is much more selective and purposeful than simply reading a book from cover to cover. You may, of course, wish to do this; however, the time available to you could make this difficult!

As mentioned in Chapter 7, for all of you as students of Criminology there is a requirement to access and read extensive amounts of text. There is also a need to 'read well' – by this we mean that you should gain some meaning from the text rather than simply read the words (Fairbairn and Winch, 1996: 8). However, different disciplines within the social sciences use particular ways of talking about their subject area and you can feel at a loss if you are not familiar with such jargon. Reading criminological literature would help familiarise you with this terminology and some social sciences have a specific dictionary that is subject dedicated; for example *The Sage Dictionary of Criminology* provides accessible definitions of terms that are commonly used in Criminology (McLaughlin and Muncie, 2001b).

To gain meaning from the literature requires you to make sure that it is relevant to and appropriate for the purpose for which you are reading. For example, no matter how interesting a book about serial killers might be, it will not help you to answer a question that concerns police powers and accountability! It is a good idea to look at the contents page and the subject index at the back of the book to find out if the book contains the information you require. From the question you should be able to identify some key words and themes. Do not forget that lectures and seminars will also direct you to relevant reading. In the first instance, it is not wise to gather as many books as you can fit into your holdall all at the same time; try to be selective. Choose two or three texts and using the contents page and the index you will be able to identify the areas of the books that focus on the topic you are interested in. This should enable you to start making a plan and your own reading list. Reading the bibliography will also help as other relevant literature will be cited. For journals, as Chapter 7 states, reading abstracts would be a useful way of gathering information and many journals are now available on the Internet where you can quickly type in key words to help you to search. Identifying key words will show you whether the article is relevant to your essay.

Remember to focus upon the specific essay title as you survey the literature. Bear in mind what the article is about, what you want to find out and if the text contains the information you need (Fairbairn and Winch, 1996: 21).

8.4   Taking notes from the literature

As you begin your reading you will need to take down notes. There are a number of ways you can do this and as you progress through university you will develop your own way of making notes. However, there are some basic points to be made about taking notes from books and articles:

- **When using books either your own or those from libraries, it is *not* a good idea to write in them or highlight areas you find interesting or useful for your essay. Books are precious commodities within academia and are used by many students and lecturers. If you have taken or been given photocopies of articles to read you** can write comments and notations on them, but, if the textbook is used by others and not owned by yourself, you should not write anything in it.

- A large writing pad with a left hand margin is useful. Write full details of the book or article at the top of the page; that is, the author's name, year of publication, title, publisher and place of publication.

- When you locate information that is useful make notes about it and put the page number you have taken it from in the margin or within the text itself. This makes it easier to locate and reference when you come to write your essay.

- If you are taking direct quotes from the literature you should make sure this is clear and identify the page number.

- Good notes will be clear, brief, easy to refer back to, include references and be well organised.

Getting into the habit of writing notes while undertaking your reading is a very useful habit since reading and perhaps re-reading a section of a book and writing your own notes on this will aid your understanding of the subject area. The purpose is not only to identify and map out the information but also to understand it; putting the material into your own words enables you to gain a much deeper comprehension of the subject area.

KEY TIP

You should aim for no more than 10 per cent of your essay word length as quotations, otherwise your work may appear as a list of someone else's ideas rather than your own.

USING THE INTERNET

Locating information on the Internet using key words is a good way to find articles about your subject area; however, it is unlikely that you would be able to create a rounded discussion that is underpinned by a sound theoretical and critical knowledge base from this alone. Websites are important sources of information for criminologists, (see section 7.4) particularly in terms of accessing current policy issues, new legislation and reports as well as journals, newspapers and discussion forums, but information can remain on the Internet for many years and become dated.

COLLECTING DATA (PRIMARY AND SECONDARY SOURCES)

Data collected are usually referred to as primary and secondary sources. Primary sources are an author's own ideas and work, for example, Karl Marx's *Das Kapital* or Foucault's *Discipline and Punish*, while secondary sources refer to an author's interpretations of other people's work, for example, Hazel Croall (1998: 269) considers Edwin Sutherland's definition of white collar crime. Thus, if you are looking at Sutherland *in* Croall you are using a secondary source, whereas if you took out Sutherland's work to consider his definition this would be using a primary source.

KEY TIPS

Keep your essay title in mind as you read the literature. Ask yourself whether or not your question is being answered.

Do not copy huge chunks from books and articles. Try to highlight the key points and gather useful quotations to provide evidence for your argument.

Make sure quotations used are clearly distinguishable from your own words.

8.5   Writing the essay

After gathering your information the next stage is to structure your writing. Regardless of whether you are constructing a report, an essay, a response to an examination question or a presentation, the most basic requirement of any writing usually involves an introduction to the topic, a discussion in the middle and an end with some sort of conclusion that has been reached. It might seem obvious to state that every essay should have a beginning, middle and an end, but this does not always follow in practice! For most of us the middle part of an essay is often the easiest to construct as this is where the main discussion takes place; however, introducing your topic so that the reader knows what to expect is an essential element of good writing.

Without a conclusion the essay is left hanging in the air and the sense of your argument becomes lost. Planning and structuring essays is therefore necessary if the essay is to communicate effectively your understanding of the question.

KEY TIP

Always keep several back-up copies of your work; disks and computers have a nasty habit of getting mangled, breaking down, crashing or blowing up on or near submission dates. (This also happens to lecturers!)

Essays should contain these four main components:

1 **an introduction**

2 **main body**

3 **conclusion**

4 **bibliography.**

INTRODUCTION

You have focused on the question to make sure you understand what the question wants you to do and how you intend to answer it. In your introduction think about relating how you will answer the question to the reader of your essay. An introduction should set out the main points of your discussion. You should also provide definitions and clarifications in your introduction. If, for example, your essay title uses the word poverty in relation to crime causation, it would be useful to get different definitions of poverty from the literature to provide an understanding of absolute and relative poverty and then move on to discuss the links between this and crime causation.

MAIN BODY

In your main body try to think of each point as a paragraph – this will add structure to your essay and will make you think about what you think the six or so main points are. Think of examples that could illustrate your points and use quotes, themes and perspectives and other research and relevant information from the literature to support your argument.

As you proceed make sure you cite references including page numbers where necessary. If quotations used are more than two or three lines long they should be inset with quotation marks at the beginning and the end of the text. The reference should be placed directly underneath the quotation. Try to make your discussion flow; this can be difficult sometimes as complex arguments often need longer paragraphs to make key points. Short bite-size paragraphs of two or three lines should be avoided as discussion appears stilted and unconnected and this would lose you valuable marks so try to use links between paragraphs to maintain the coherence of the argument.

CONCLUSION

In your conclusion draw the threads of the discussion together and reaffirm your main argument. Make sure your conclusion explicitly addresses the essay title. At this stage take care to focus on what you have discussed so that you do not introduce another aspect of the topic into the conclusion.

KEY TIP

It is often very helpful to read your essay aloud at this point since quite often you can get too used to looking at your own work and reading aloud helps you to hear if the sentences are constructed badly or do not make sense. You can also judge whether your argument is relevant and makes sense. This is often best done after taking a break from your work; after you have relaxed you can return to your work with a fresh pair of eyes.

8.6 References and bibliography

The notes about how and when to use references within this section will apply to all of the academic work that you undertake. Harvard references are the preferred model and they are also a clear way of showing the reader of your work the sources that you have used in producing your essay, presentation or report.

KEY TIP

If you forget how your bibliography should look in terms of layout simply have a look at the bibliography in the back of a Criminology textbook to check you have noted all the bibliographic details.

There are several reasons for using references:

- It avoids claims of plagiarism that could be brought against any student who uses others' words, ideas or phrases without attributing them to those who coined those phrases. In academia this is similar to stealing and is looked upon as being a serious offence of cheating.

- References also allow the reader to see how much research has been undertaken in producing a piece of work and what the quality and range of these sources are.

- References also allow the reader to check that the quotation used and attributed to a certain author in a certain year was actually used by that author in that year.

Consider the examples given below:

> In any case, Visher (1987) calculated that a doubling of the prison population during the 1970s achieved only a small decrease (6%–9%) in crime rates for robbery and burglary. For the early 1980s, imprisonment was 'only slightly more effective in averting crimes'. (p. 519)

> Taken from Lilly, Cullen and Ball (2002: 228).

The author's name and year of publication are clearly referenced and a page number for the direct quotation is given so that the reader can readily access and check the validity of the reference given.

The next example is a little different:

> There is little empirical evidence for consensus theory's claims about the desirability of self-regulation (Dalton, 2000; Dawson et al., 1988; Smith and Tombs, 1995). The case against strict enforcement regulation is always made

hypothetically, since it has never been tried and tested over a sustained period (although see Alvesalo, 2003a and 2003b).

Taken from Whyte in Muncie and Wilson (2004: 140).

If we take the excerpt above and look at the way references are used within it we can see that Dalton, Dawson, and Smith and Tombs are all cited in the Harvard style and the year relates to the year the book or article was published. In this way Dalton 2000 relates to a book or article written by Dalton and published in 2000. Alvesalo 2003a and 2003b relate to two pieces of work by an author called Alvesalo which were both published in the same year and a and b differentiate these.

For direct quotations the page number follows the author's name and year of publication. Quotation marks must be cited when using direct quotations to clearly show which are the author's words and which are yours. This is shown again in the example below:

Braithwaite is in favour of compliance strategies, but only when they are backed up by tough sanctions: 'regulators will be able to speak softly when they carry big sticks' (Braithwaite, 2000: 99).

The above excerpt was also taken from Whyte in Muncie and Wilson (2004: 140). From this reference the reader of your work can see that you are quoting Braithwaite's publication of 2000 and if necessary can go to page 99 to examine the quotation.

For ideas, quotations, words, phrases, historical dates or events a reference is required within your work and this needs to be supplemented at the end of your work with a bibliography. The bibliography should be in alphabetical order of author's surname. Using the references from within the excerpts above the bibliography would look like this:

Alvesalo, A. (2003a) *The Dynamics of Economic Crime Control*, Vol. 14. Espoo, Finland: Poliisiammattikorkeakoulun tutkimuksia.

Alvesalo, A. (2003b) 'Economic crime investigators at work', *Policing and Society*, Vol. 13 (2), pp. 115–38.

Braithwaite, J. (2000) *Regulation, Crime, Freedom*. Aldershot: Ashgate cited in Whyte Chapter year pp in Muncie and Wilson (eds) ————— (2004).

Dalton, A. (2000) *Consensus Kills: Health and Safety Tripartism a Hazard to Workers' Health?* London: AJP Dalton.

Dawson, S., Willman, P., Bamford, M. and Clinton, A. (1988) *Safety at Work: The Limits of Self-Regulation.* Cambridge: Cambridge University Press.

Lilly, J.R., Cullen, F.T. and Ball, R.A. (2002) *Criminological Theory: Context and Consequences*, 3rd edn. London: Sage.

Muncie, J. and Wilson, D. (2004) *Student Handbook of Criminal Justice and Criminology.* Cavendish: London.

Smith, D. and Tombs, S. (1995) 'Beyond self-regulation: towards a critique of self-regulation as a control strategy for hazardous activities', *Journal of Management Studies*, Vol. 35 (5), pp. 619–36.

REFERENCE STYLES

For an article/newspaper:

The author, (year) 'title', *Journal Title*, Vol. (issue), pages.

The author's surname would come first followed by initials and then the date of publication in brackets. The title for an article would be in inverted commas and it would be followed by the title of the journal or newspaper, which would be in italics or underlined. This would be followed by the Volume number and the issue would be in brackets and lastly the pages used would be cited. For example:

Smith, A. (2004) 'Crime and the Underclass', *Prison Works*, Volume 4 (6), pp. 136–180.

For a book:

The author, (year) *Title*, place of publication, publisher.

The author's surname would come first followed by initials and then the date of publication in brackets. The title of the book would then follow and would be written in italics or underlined. This would be followed by the place of publication and the name of the publisher. For example:

Brown, D.A. (2004) *What About Corporate Crime?* London: Sage.

For a reference to a chapter in a book that has
been edited by one or more people:

The author (year) 'Title of chapter' in editor's names (eds) *Title of book*, place of
   publication and publisher.

The author's surname would be followed by the date of publication in brackets. The
title of the chapter would be in inverted commas, followed by the editors' names and
(eds) in brackets. The title of the book would be written in italic or underlined next
followed by the place of publication and the name of the publisher. For example:

Saraga, E. (2001) 'Dangerous Places: The Family as a Site of Crime', in Muncie, J.
   and McLaughlin, E. (eds), *The Problem of Crime*. London: Sage and Open
   University.

For a website:

If there is an author's name on the website it would be cited and then followed by the
full URL – which is the exact address of where you found what you are referring to.
For example:

http://www.homeoffice.gov.uk

The date that you accessed the website should follow this in brackets.

Activity 8d: Constructing
a bibliography

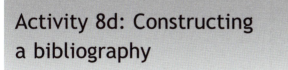

Write out a bibliography using four books, three journal articles and three
websites in the form shown above.

Finally if we consider the introduction to your essay and the way we use the work of others, as discussed above, to set the context of your work, the following three extracts clearly demonstrate the correct way in which to present your work. Given below are three examples of essay writing styles, all deal with the same subject matter but one is the basis for a good essay, one is poor and may even fail, while the third is plagiarised. While your own writing style may be different to these examples, the underlying principles of how to present academic essays should be evident to you.

Question: Was the development of the 'new police' a rational social development or an attempt to socially control the urban working classes? Critically evaluate both viewpoints.

ANSWER 1

Introduction:

The term 'new police' is used to describe the police forces that emerged during the early nineteenth century. Starting with the establishment of the Metropolitan Police in 1829, police forces across England and Wales began to be established. There is no argument about the chronological facts but there is a debate about why the 'new police' emerged at this time. There are two competing histories of the police. The first, often described as the 'orthodox' history of the police, saw the development of policing as progressive and rational as the local, less formal systems of policing that had previously existed were no longer able to cope with rising levels of crime and disorder (Reith, 1943). There is an alternative perspective and this suggests that the 'new police' emerge as a tool to be used in controlling the growing urban working class (Storch, 1975). In order to answer the above question it is necessary to discuss the strengths and weaknesses of the two competing histories and to provide a critical evaluation of their merits. Another significant factor here is the importance of social and economic change during this period that saw a rise in the urbanisation of society and a reduction in the rural economy.

ANSWER 2

Introduction:

The new police were established by Robert Peel in 1829. They were sometimes known as the blue locusts or peelers. The new police came about because of the industrial revolution and because people were living in cities the new police were needed to deal with all the crime. Reith said that the new police were the answer to crime. Storch said the new police were used to control the working class who were committing crimes like gambling, drinking and prostitution.

ANSWER 3

Introduction:

There is not one school of police history; the divisions between the different views of the origins of the police in Britain tend to reflect the way in which different people view the police and their role in society. Storch (1975) and Reith (1943) are seen as significant contributors to this debate, representing the revisionist and orthodox views respectively. The orthodox interpretation argues that the Metropolitan Police were such a success that the model was introduced throughout the country. Storch argues that the police were needed to control the new urban population, and the police were seen as an institution established as another element in a broad strategy of control.

All three introductions are dealing with the same points:

- **In the first there is an attempt to identify the key debates and point to relevant literature and assuming the essay continues in the same manner then this essay should receive a good grade. The writer has attempted to demonstrate *her/his own* understanding of the question but intends to refer to the work of others to support the discussion.**

- **In the second example the writer is expressing the results of her/his reading but not in a coherent style. The points are not well structured, referencing is poor and it is unlikely that an essay continuing like this would be awarded a high mark.**

- **At face value the third example appears to be well written and references relevant sources. However, the work has been plagiarised and to an academic this would be quite evident. The essay if it continued like this would at least fail and, depending on the extent of the plagiarism throughout, harsher sanctions could be imposed. (The text has been plagiarised from McLaughlin, E. and Muncie, J. (eds) (1996) *Controlling Crime*. London: Sage, pp. 8–14.)**

REFERENCES

McLaughlin, E. and Muncie, J. (eds) (1996) *Controlling Crime*. London: Sage.

Reith, C. (1943) *A Short History of the Police*. Oxford: Oxford University Press.

Storch, R. (1975) '"The Plague of Blue Locusts" Police Reform and Popular Resistance in Northern England 1840–1857', XX *International Review of Social History*, 61.

8.7 Presentation

Clear and neat presentation of all of your work is very important. Your work will need to be word processed before submission (except in the case of examinations) so you must make sure that you add some time into the timetable for typing your work up. You will usually find that lecturers and tutors prefer you to leave a left and right hand margin and like to see work double-spaced so that when it comes to marking your work they have space to write comments onto your work. Check with your tutors and module guides/handbooks to see how they prefer your work to be presented – whether they have a standard expectation that all of your work will be in Arial font size 12 or Times Roman 14 and double line spaced, for example. Using a word processor is useful in terms of changing the layout or order of your work easily and there is a spell-check – though you really need to be careful and check that this is set to UK rather than US spellings. You will still need to supplement this with proofreading to flag up grammatical and spelling errors.

8.8 Marking criteria

Many lecturers now provide a copy of how your work will be marked in their module guides or on the intranet or blackboard site. These are sites that use Information Communication Technologies and are usually only accessible to staff and students within the university. They are sometimes known as Virtual Learning Environments. Using this information can be very useful because it allows you to identify what academic standards are expected by the lecturers who are setting and marking your assessments. This will provide a guide for you to follow and you can use this as a kind of check list to see if your work meets the criteria set before submission.

An example might include the marking criteria for essays and this may be separated into three distinct sections:

1 **Presentation: this refers to grammar, spelling, structure of the essay, writing style**

2 **References and Bibliography: this shows you have illustrated research, knowledge of the subject area, contemporary material, books, journals**

3 **Content: this refers to critical thinking, argument's sense, coherence, understanding of the topic.**

If your lecturer or tutor does provide this information for you then check your work against this before submission in order to gain as high a grade as possible. (See Appendix 5 for an example of an essay mark sheet.)

8.9 Dissertation writing

Your dissertation is a piece of work to be submitted within the final year of your undergraduate Criminology degree. This is the longest, most substantial piece of work that you will undertake and it requires forward planning in order to go as smoothly as possible. Finding the subject for your dissertation is sometimes the most difficult element of starting to think about this piece of work, though it is worth some real reflection upon which subject areas you have found the most interesting within the first two years of your degree programme. Finding a subject that will maintain your interest and enthusiasm is important since the length of this piece of work is usually up to 10,000 words in length and in some instances can be longer. Some of you may begin to think about a subject area that you would like to investigate further when you are in the second year of your degree. If you can choose the area before heading off for the summer break then you can ask interested lecturers for their opinions on the subject you are considering and might get an indication of who may supervise your work. You would also be able to use the time during the summer to do some research for relevant literature on the subject that you are interested in. This allows you to do some of the background research before you are required to work on assessment for the third and final year of your degree.

Here are some general things to check:

- Check when your dissertation registration form has to be submitted.

- Begin background reading on the subject you are interested in writing a dissertation on — this will allow you to develop a research question/title for your dissertation.

- Check which member of staff will supervise your dissertation.

- Begin researching and drafting your dissertation — cut it down into smaller parts by having an introduction, 2–3 chapter areas, and a conclusion.

- Ethics — before you begin undertaking the research for your dissertation you must complete an ethics form and have this checked before your dissertation can commence.

- Check that your research takes ethics into account by checking the British Society of Criminology Ethical Code (www.britsoccrim.org/ethics).

- Remember to leave enough time for gathering sources — some of the books you will want to use may need to be sent for from other libraries using an inter-library loan form — this process might take a while.

- Draft your literature review before moving on to the other chapters.

- Write up your dissertation, checking and gaining feedback from your supervisor regularly.

- Make sure your work conforms to the expected presentation and have your dissertation bound.

- Check final submission dates for your dissertation and submit in time.

KEY TIPS

Make sure you keep your dissertation on a floppy disk, the hard drive of your computer and also print out a hard copy to cover computer failure or other unforeseen circumstances.

You may choose to have the dissertation typed up professionally. This often means that you present your work to a high standard – check notice boards within the university to see if this service is advertised.

Get hold of a dissertation guide book, or the guidelines used from your university so that you are aware of how your dissertation should be presented.

Ask your dissertation tutor where past dissertations are held and go and have a look at them.

8.10   Feedback

Feedback from your essay writing and from all types of assessed work is useful for all students. From feedback you not only gain a mark/grade for your work you also receive an indication of your strengths and weaknesses so when above we considered a mark sheet which had been given three sections to use as a check list, you will gain an understanding of how well you have performed in relation to the three sections – so you will see how well you did in terms of presentation, references and bibliography, and content. This information will illustrate both strengths and weaknesses and will provide you with the information that you will need in order to reflect on whether you need to do further work and where this further effort is required as well as signalling your strong areas which can be reinforced. It is, therefore, important to collect the feedback that you receive and read through it carefully, and if you are in doubt about how you can improve on weak areas then booking a tutorial session to go through the feedback with the lecturer who marked your work is also very useful.

8.11   Summary

- **This chapter has attempted to provide a guide for you when faced with writing your essays including preparation for this, note-taking, data collection and using** references as well as guidance on undertaking a presentation, constructing a bibliography, considering marking criteria and providing notes on dissertation writing.

- This guidance will be useful throughout your degree course at all levels and may prove useful beyond your degree and into the world of work. Skills such as presenting your ideas and constructing an argument remain useful inside and outside formal education.

9 Criminological Theory and Criminal Justice Policy and Practice

CHAPTER OVERVIEW

By the end of this chapter you should be familiar with:

➤ criminological theories and their application to policy and practice
➤ the role of the criminal justice agencies and their responses to crime and disorder.

This chapter is designed to introduce you to some criminological theories and to help you to develop skills in analysis, critique and the application of theory to practice. This will enable you to assess the merits and weaknesses of a range of theoretical perspectives. It is important to remember that responses to crime are inextricably linked to the way in which policy makers explain criminal behaviour, and to be successful in your degree you must demonstrate that you understand how these matters interact and relate to each other. Criminological theories endeavour to explain the causes of crime and criminality and also inquire into the ways in which society attempts to control crime. As discussed in Activity 9g in this chapter some criminological theories seek to explain crime as a social construction. This means that the interaction between different groups in society leads to some behaviours being defined as illegal, criminal or anti-social at particular times in history. Others focus much more on individual behaviour and some concentrate on biological and psychological explanations of criminal activity. It is important to recognise that these theoretical perspectives are often in conflict with each other. Indeed, many of the current debates about crime and disorder are based on a variety of these theoretical positions. They are evident in many of the policy decisions that have been discussed over recent years and they frequently appear to focus on those perspectives that are concerned with individual responsibility (for example, parental responsibility), association with other offenders (for example, anti-social behaviour) and social exclusion (for example, concerns about the 'underclass'). Some theories will be easier to accept than others and it is important that you keep an open mind when studying them because our own life experiences, cultural, religious and political beliefs will influence our decisions about which theoretical perspectives appear to be the most appropriate to inform our understanding of crime and criminality.

Criminological theory has an important impact on the way in which criminal justice agencies respond to crime and disorder and it also affects significantly the way in which criminal justice policy is developed. A key skill is analysing new and existing criminal justice legislation and making connections between the theories that underpin debates that focus on crime and the responses to crime that are supported by policy makers. The activities in this chapter focus on issues that are currently considered to be important in understanding crime and disorder and anti-social behaviour. Working through these activities will allow you to consider the brief outlines of some key theoretical perspectives while you apply this to some contemporary debates surrounding specific policy issues. In this chapter we will also seek to expand on the links between theory and practice, and again the activities encourage you to make these connections.

First, it is necessary for us to consider some of the major theoretical debates that have dominated Criminology over the past century as these have been a major influence on criminal justice legislation throughout the twentieth century. These debates have involved discussion about the value of punishment over rehabilitation, the concept that prison works, whether a more punitive response or a more inclusive response will have the greatest impact on crime, and at other times whether crime was a consequence of structural inequalities in society or could be attributed to biological or psychological factors. Clearly, whichever view dominates at any time, the resulting legislation will reflect the theoretical stance. It may be useful to begin with an activity to establish your own perspective on crime before we go any further!

Activity 9a

Consider the following statements and say as honestly as you can whether you agree or disagree with them:

- Criminals commit crime because they are bad people; they have flaws in their character and are essentially evil individuals.
- Criminals commit crime because they are excluded from mainstream society and are unable to benefit from the things that others in society enjoy.

- The easiest way to reduce crime is to lock all criminals up for as long as possible.
- The demands of modern society encourage some individuals to behave in ways that are defined as criminal.
- Different rules apply to the rich and the poor as far as crime and disorder is concerned.
- The lower classes commit most of the crime in most societies.
- There are some offences or types of behaviour that are seen as criminal in all societies.

Keep your answers by you and see which of the following theories applies to the different statements above.

Criminological theory is not a theoretical tradition that stands alone, it is based on a number of other traditions and perspectives. As mentioned earlier, there are several areas of the social sciences that make major contributions to our understanding of crime and indeed some of the earlier traditions are from the natural sciences. For example, the biologically based theories were developed in a medical context (see Lombroso (Williams, 2004)) and to some extent the 'medicalisation' of criminality continues to be significant especially with advances in our understanding of genetic sciences. One example could be the use of genetic research that is seen as an attempt to resolve the nature versus nurture debate (Williams, 2004). A word of caution! It is important to consider the strengths and weaknesses of all theoretical perspectives and to recognise that in various circumstances they may all have some merit; as with all research and science there will be ideas and perspectives that are dominant in one era but that lose their popularity or indeed credibility in another.

9.1 Criminological theories

Generally speaking criminological theories can be divided into several different categories:

BIOLOGICAL PERSPECTIVES AND PSYCHOLOGICAL PERSPECTIVES

The focus of these theories is that criminals are born with a certain predisposition to offend. This is because of their inherited genetic or physiological make-up. While environmental factors are not excluded from these theories, these social factors are sometimes seen as the triggers that lead to the individual committing crimes, although biology or psychological factors are seen to predispose the individual to offending. Early writers such as Lombroso and Ferri were interested in the physical appearance of criminals. It is suggested that theories that focus upon inherited genes as indicators of criminal behaviour are better indicators than social factors when predicting criminality. However, these theories are frequently developed by focusing on those who are convicted of crimes; this ignores those who are seen as law abiding and who share similar characteristics, and consequently fails to explain variations in behaviour between similar biological or psychological types. Nevertheless, there are many theoretical approaches developed from the biological perspective with writers such as Herrnstein and Murray (1996) who explore IQ as a factor in criminal behaviour.

SOCIOLOGICAL PERSPECTIVES

By contrast, social learning theory (Bandura, 1977; 1986) was an attempt to identify links between observable behaviour and the environment. Social Disorganisation perspectives seek to make connections between environmental factors and the commission of crimes. This is a theoretical position that is linked to the work of the '**Chicago School** of Sociology' (see Box 9a).

Box 9a The Chicago School

Based on the work Robert Park and his colleagues (1925) this perspective examines the 'social ecology of the city' (McLaughlin and Muncie, 2001b). The focus is on the way in which modern cities expand from their centre with the inner city representing the run-down, transient areas that encircle the business district. Here, there are high levels of poverty, ill health and deprivation. The working-class and then the middle-class districts encircle this 'zone' before reaching the affluent outer district. It was in the dilapidated inner city areas that the Chicago School sought to identify the causes of crime and **deviance**. They attempted to do this by an innovative research methodology involving **participant observation** and focused interviews. This involved observing and recording the daily lives of the inhabitants of this part of the city.

A dictionary definition and discussion of the Chicago School is provided by Keith Hayward (2001) in McLaughlin and Muncie (2001b).

Another theoretical perspective that has its roots firmly in the discipline of Sociology is '**anomie**' developed initially by Emile Durkheim in the late nineteenth century and further explored by Merton in the 1930s. This can be defined as 'a state of ethical normlessness or deregulation, of an individual or society' (Morrison in McLaughlin and Muncie, 2001b: xx). One key idea of this perspective is the assumption that we all share the same goals, such as being successful, creating wealth or consuming goods. For Merton, the American Dream of 'success', was expressed as 'money success'. The desire to be successful in this context was seen as a legitimate goal, however, for those unable to attain this goal legitimately; illegal activity was seen as the means to the end. This results in anomie. One of the key theoretical perspectives aligned to anomie is Merton's (1938) Strain Theory (see Box 9b).

Box 9b Strain Theory

The argument here is that people tend to commit crime when they are unable to achieve their goals legitimately. Frustration and anger set in and they attempt to get what they want through:

- criminal activity, such as theft, burglary or fraud
- or, they become angry and express this through violence towards other people
- or, they take refuge in escapism, such as alcohol abuse, drug-taking or self-exclusion from mainstream society.

Strain Theory seeks to identify the stresses that lead to strain and whether these strains lead to decisions to commit crime.

Control Perspectives are another theoretical stance based on sociological approaches. Key here is the notion of conformity and the assumption that human nature is based on independence and free will, which leads to non-conformity. Here, the focus is on the ability of institutions (for example, parents, peers, schools, spouses and jobs) to influence human behaviour. The key emphasis is on the strength or weakness of ties to society; hence, those with weak ties (socially excluded) may feel less need to conform than those with strong ties (socially included).

Labelling/Interactionist Perspectives focus on the way in which one group in society identifies the behaviour of others as problematic (see Box 9c). This leads to those whose behaviour is deemed to be unacceptable being defined as troublemakers, criminals or anti-social. This perspective is discussed further in Activity 9g later in this chapter that looks at the social creation of crime.

Box 9c Labelling Theory

This relates to the process by which certain groups categorise other groups in society. In the main, this refers to the way in which the state, powerful groups and the law classifies the behaviour of the less powerful groups in society. Individuals and groups are labelled or stereotyped as criminal, deviant or anti-social and they may respond to the label. Consequently, they could adopt the form of behaviour that the label was intended to prevent.

Conflict Theory (see Box 9d) and consensus theory are seen as two opposing perspectives. Consensus is based on the assumption that society has a common acceptance and agreement of right and wrong and that this is the basis of social stability. By contrast, conflict theory sees society as a collection of different groups who oppose the consensus or who challenge the rule of law because they feel it is unjust. This could be the case when those with the power to legislate against certain behaviours are seen to have gained that power illegitimately.

Box 9d Conflict Theory

This theory is often contrasted with the positivist view that there is a consensus in society. It involves conflict based on group identity, class identity or cultural identity. It is often seen as relating to power and authority and attempts to understand the way in which the criminal law serves the interests of specific groups in society.

FEMINIST PERSPECTIVES

Feminist Criminologies challenge the male-centredness of criminological study. They can be seen to come from two main traditions. Firstly, liberal feminist criminologies that see men and women as essentially the same with women being denied the opportunity to engage in the same activities as men, including crime (Daly, 2001). Secondly, theories developed from critical social theories that focus on gender power relationships rather than role differences. It is worth noting that feminist criminologies are relatively new having emerged towards the end of the twentieth century. Equally there is no one feminist perspective as they make diverse claims.

It is not possible in this text to discuss all theoretical perspectives in depth. There is a wealth of texts that do that and the examples given above are an indication of the variety that exists. Further indicative reading in relation to theoretical criminology is included at the end of this chapter.

Hopefully by now you will have appreciated the importance of assessing any criminological topic from a variety of viewpoints and the need to apply these perspectives to any assessments and essays that you are required to undertake. Activity 9g below is a good example of how you can use a theoretical perspective to look at a variety of situations.

9.2   Criminology theories, policy and practice

In this chapter we now intend to introduce some examples of these theories and examine the ways in which they have been used to develop criminal justice policy and how they inform the practices of criminal justice agencies. It should be noted that some writers (Downes and Rock, 1998: 326) suggest that the academic study (theory) of the discipline is often seen not to be contributing sufficiently to policy development. We, therefore, need to consider to what extent this actually happens. If we return briefly to the extract in Chapter 2 from Muncie and McLaughlin (2001: 15–16) we could argue that supporters of the notion that crime is socially constructed would claim that policy makers have the greatest influence over the existence of crime. For example, by legislating for the decriminalisation of certain behaviours we can stop that behaviour from being represented as a criminal act. Recent debates concerning the decriminalisation of certain types of drug use would be one example of this; equally, the current debates relating to the prohibition of smoking in public places could see such an activity that is currently legal becoming 'criminalised'.

The following key points drawn from debates by MPs in relation to Britain's drug laws suggested that they should be radically amended.

- **The Home Affairs committee recommended that ecstasy should become a class B drug. This would put it on the same level as the current classification for amphetamines.**

- The reclassification rather than decriminalisation would reduce the maximum sentence for those found carrying the dance drug from seven years to five.

- It was also suggested that cannabis should be made a class C drug and this was made policy during 2004.

- They also recommended trials of heroin prescription programmes for addicts and the provision of safe injecting rooms for heroin users.

This type of debate reflects the theoretical debates surrounding society's reaction to offending behaviour. Is it best to punish offenders in an attempt to make them change their behaviour or should we offer treatment to help them to address their drug habits? This notion of treatment versus punishment can be seen to relate to discussions about whether criminals are born or made, the nature–nurture debate.

As criminologists studying both theory and criminal justice policy it is our intention to explain the development of legislation and indeed the introduction of new legislation by drawing attention to the underlying ideology and theoretical concepts that influence such change. Making an analysis in this way is therefore a key skill that you would be expected to develop as a student on a Criminology degree programme. In doing this you might argue that the increased emphasis on custodial sentences reflects a belief that criminals are essentially 'bad' and as a consequence should be excluded from the rest of society who are 'decent, law abiding individuals'. This view could be linked to some of the biological or psychological theories that exist. Such theories would argue that 'all behaviour is motivated and purposive' (Lilly et al., 2002: 25). This contrasts with the notion that 'crime' is the consequence of our interpretation of behaviour as criminal, as labelling theorists suggest (Lilley et al., 2002: 105–25). To pursue the example of an increase in the prison population, labelling theorists would argue that the behaviours of a particular group in society were defined as problematic enough to warrant a custodial sentence. Perhaps the high proportion of those in prison with a history of substance misuse may be attributed to the state's view that such behaviour is unacceptable and their consequent attempt to reduce these activities results in a higher prison population.

You will have now recognised the numerous debates within Criminology that continually throw up conflicting positions: for example, deterrence versus rehabilitation in relation to punishment, or punishment versus welfare, which is another way of describing this; or, disorderly youth versus disorganised communities which would be a feature of the theories developed by the Chicago School. Such dualities are evident throughout criminological literature and it is important that you consider the merits and disadvantages of each approach when studying Criminology. This is a skill that often requires you to challenge your own subjective views as you attempt to analyse objectively the views expressed by others. Should you wish to challenge a particular perspective you need to use the skills discussed earlier in this book in relation to supporting your argument with evidence rather that expressing your own opinions.

This chapter will now provide an opportunity to look at some specific examples of recent policy and asks you to consider which theoretical perspectives have contributed to the development of policy or influenced the debates surrounding the issues raised. The first example, focusing on Parenting Orders, will initially provide a brief overview of this type of order before going on to discuss some related theories. Clearly, there are many more theories that relate to Parenting Orders than we examine here, and this is something that you may go on to reflect upon as you progress through your degree programme. For some of you this may appear to be a rather complex example, but as your own learning progresses you should be able to deconstruct the various strands evident in this discussion.

PARENTING ORDERS

In 1998 the Crime and Disorder Act received Royal Assent and part of this legislation included the introduction of the Parenting Orders. The Parenting Orders, which commenced in June 2000, require parents/carers of convicted young people to attend counselling and guidance classes for no longer than three months. However, the Court may impose a second element which requires parents/carers to exercise control over their child's behaviour and to comply with particular requirements, for example, ensuring that their child attends school; this may last twelve months. A failure to fulfil the conditions can be treated as a criminal offence and the parent/carer can be prosecuted. According to the *Observer*, 'The sanctions for parents who failed to comply were designed to be a tough deterrent: a fine of up to £1,000 or a prison sentence of up to six months' (2004). As noted by Goldson and Jamieson, 'on the one hand the Orders convey the language of "support", on the other they consolidate and extend punitive measures' (2002: 89). This can therefore be said to be a somewhat ambiguous piece of legislation. Thus Parenting Orders have not been without critics and the dichotomy of opinion is largely related to which particular theoretical stance is supported. Broadly speaking, on the one hand there are those that argue that crime is largely located in and around the individual (individual agency) focusing upon 'pathological' family forms (Goldson and Jamieson, 2002) and, on the other, those that argue that it is about broader social situations and opportunities, largely tied to poverty (social structures), and then there are those that take a mixed view that it is a combination of both.

There are numerous criminological theories that can be related to Parenting Orders and within this chapter we will examine a number of these to illustrate how theory can be linked to practice, starting with the work of Farrington (1992).

In 1961 a sample of 411 working class boys aged eight were selected from six schools in London for a study on **delinquency**. This study is well known in Criminology and is referred to as the Cambridge Study in Delinquent Development. They were contacted several times over the duration of the study when they were aged 10, 14, 16, 18, 21, 25, 32 and 46 to explore which of them had developed a 'delinquent way of life' and to examine reasons why this had taken place. (For more information on this study see Muncie, 2004: 26). Farrington and colleagues have, over the years, developed a set of risk factors that they believe increase the chances of a young person turning delinquent. These have been summarised by Muncie as:

- Socio-economic deprivation (e.g. low family income/poor housing);
- Poor parenting and family conflict;
- Criminal and anti-social families;
- Low intelligence and school failure;
- Hyperactivity/impulsivity/attention deficiency;
- Anti-social behaviour (e.g. heavy drinking, drug taking, promiscuous sex). (2004: 26)

This theoretical stance appears to have been hugely influential in shaping New Labour's criminal justice and in particular youth justice policy. Clearly, Parenting Orders are designed to try to combat poor parenting and family conflict, yet one needs to ask questions about what constitutes good and bad parenting and whether we all agree on such definitions.

Work such as Webster et al. (2004) has, however, argued that Farrington et al.'s model is overly deterministic and that transitions are complex and unpredictable. Farrington et al. suggest that anti-social behaviours developed in childhood become amplified in later life.

In the Cambridge Study, delinquents had a tendency to be troublesome and dishonest in their primary schools, aggressive and frequent liars at age 12–14, and at age 14 were likely to be bullies. By the age of 18, this anti-social behaviour was evident in a wide variety of respects, including heavy drinking, heavy smoking, using prohibited drugs, and heavy gambling. In addition, they tended to be sexually promiscuous, often beginning sexual intercourse under age 15, having several sexual partners by age 18, and usually having unprotected intercourse (Farrington, 1992).

Webster et al. (2004), however, highlight *unpredictable* 'critical moments' (such as the death of a parent or sibling), which often have varied consequences. For some criminals, for example, these critical moments are a catalyst to move away from crime, for others they act to deepen the individual's involvement in crime and these responses (as well as the critical moment itself) are unpredictable. Another example of the unpredictability of criminal careers is the effects of broader social changes that occur within society that cannot be predicted (particularly if researchers only focus upon the individual and their behaviour!). One example of this is de-industrialisation and its impact upon opportunities for employment (see Morris, 1995, for a case-study approach that explores the effects of economic decline in Hartlepool). During the post-war boom no one would have predicted that there would be widespread closure of traditional industries with such a devastating impact upon levels of employment. Winlow (2001), for example, illustrates how de-industrialisation impacted upon Sunderland and in ship-building alone over 34,000 jobs were lost in the period 1971–1981. This decline in traditional industry has had a massive impact upon employment opportunities, particularly for working-class young men.

A number of authors (see Goldson and Jamieson, 2002, for example) have argued that the government's reading of risk factors focuses upon failed families rather than addressing the broader range of factors. Pitts (2001) illustrates how the focus on aetiology of crime became increasingly narrow as the Labour government discussed and formulated their youth crime and youth justice policy documents after their election in 1997. It is possible to argue, therefore, that what we are left with are interventions that are overly centred on the individual and their families and have failed to address the wider issues of poverty and social exclusion. Muncie convincingly argues that:

> Early interventions have invariably become individualised and behavioural. Primary attention is paid to responding to the symptoms, rather than the causes, of young people's disaffection and dislocation. The social context of offending is bypassed. (2002: 149)

Or as Pitts notes:

> Clearly families play an important role in determining whether, or to what extent, children and young people become involved in a broad range of socially deviant and illegal behaviours. However, the primacy ascribed to the 'crimino-genic' lower class family in New Labour's youth justice strategy suffers from what Elliot Currie (1985) has termed the 'fallacy of autonomy.' This is because it ignores the relationship between socio-economic stress, neighbourhood poverty and the biographies of young offenders. (2001: 10)

This focus on the family is perhaps attributable to the influences of Neo-conservative authors who began to re-emphasise the moral aspects of social theory arguing that the root cause of problems of unemployment, crime and social disorder lies in the moral fabric of society. The work of Charles Murray (1984, 1990, 1994) became very influential in Britain as he argued that the state, through providing welfare, had created an 'underclass' population. According to Murray illegitimacy, violent crime and drop-out from the labour force were clear signs of the emergence of such as population. He argued:

> If illegitimate births are the leading indicator of an underclass and violent crime a proxy measure of its development, the definitive proof that an underclass has arrived is that large numbers of young, healthy, low-income males *choose* not to take jobs. (1990: 17; our emphasis)

Furthermore, his underclass thesis centres on the apparent rise in single mothers and the growth of illegitimate births. He argues that many of the social problems suffered within the underclass are because 'communities need families' and 'communities need fathers' (1990: 7). This thesis was adopted in the UK and Dennis and Erdos (1992) argued that it was a simple matter of 'common-sense' that crime was an inevitable by-product of the break down of the traditional family as parents avoided their responsibilities. But as noted by Muncie:

> Once again the root cause of youth crime is viewed in terms of a breakdown of morality associated with dysfunctional families and a feckless underclass. (2004: 139)

Thus others have argued against this thesis and indeed Murray's work has received almost universal criticism amongst British criminologists. Many have challenged the empirical basis (or lack of it) of much of Murray's assertions. Moreover, many see the root causes of the problems as lying not with the individuals per se but the social situation they find themselves in, i.e. socially excluded and living in poor communities (see MacDonald, 1997). The answer, therefore, does not lie in Parenting Orders but instead in providing opportunities for these marginalised individuals and improving their quality of life through better housing, income, education, quality employment and leisure.

What we have illustrated then are the ways in which theories about crime and disorder can help shape our understandings and opinions about criminal justice policy. Just using this one illustration of Parenting Orders shows the importance of theoretical knowledge in shaping our interpretation of criminal justice policy and practice. The example, however, highlights the complexities of studying human behaviour by illustrating competing explanations that seek to understand this behaviour. Therefore, you must always remember that there are no right answers. Nevertheless, you must also back up your views with evidence to support your arguments; this is discussed in greater depth in Chapter 7.

Pulling together theory and practice can help us devise critiques of government initiatives that are overly concerned with populist policies that are devised to appeal to the electorate (often punitive) rather than policies that are informed by criminological knowledge derived from empirical, academic research. It is important, therefore, that you remember to draw together a range of knowledge from your studies and do not fall into the trap of thinking that your modules or units of learning are separate entities that once you have completed them you can forget about their content. A good student will always draw on all areas from their programme rather than just the topics covered within individual modules.

FURTHER READING

Parenting Orders:

Muncie, J. (2004) *Youth and Crime*, 2nd edn. London: Sage.

Newburn, T. (2003) *Crime and Criminal Justice Policy*, 2nd edn. Harlow: Pearson Education Limited (Chapter 8: Youth Crime and Youth Justice).

www.youth-justice-board.gov.uk

Criminal careers:

Farrington, D. (2002) 'Developmental and risk-focused prevention', in M. Maguire, R. Morgan and R. Reiner (eds), *The Oxford Handbook of Criminology*, 3rd edn. Oxford: Oxford Unviersity Press.

Muncie, J. (2004) *Youth and Crime*, 2nd edn. London: Sage (Chapter 1: Youth Crime: Representations, Discourses and Data).

Underclass and social exclusion:

MacDonald, R. (ed.) (1997) *Youth, the 'Underclass' and Social Exclusion*. London: Routledge.

Muncie, J. (2004) *Youth and Crime*, 2nd edn. London: Sage (Chapter 4: Explaining Youth Crime II: Radical and Realist Criminologies, and Chapter 6: Youth and Social Policy: Control, Regulation and Governance).

Activity 9b

Using the following two quotations, consider how social control theories relate to Parenting Orders.

> The introduction of additional sanctions for parents and children appears to represent a further tightening of control. (Smith, 2003: 69)

> What is noticeable in this gamut of legislative reform is that virtually any intervention, monitoring and scrutiny of young people's lives can be justified in the name of crime prevention. (Muncie, 2002: 151)

Activity 9c

From your reading on criminological theory, relate other relevant theories to Parenting Orders.

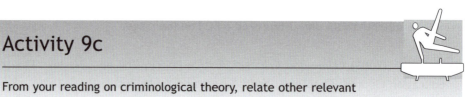

Activity 9d

Now we have explored the example of Parenting Orders and related theories, we would now like you to explore the connection between criminological theory and the following pieces of criminal justice legislation:

Referral Orders

Part 1 of the Youth Justice and Criminal Evidence Act 1999 introduced the Referral Order. These Orders are almost mandatory for 10–17 year-olds pleading guilty at a first Court appearance. The only other sentencing options are Custodial Sentence (if the offence merits it), an Absolute Discharge or a Hospital Order (Earle et al., 2003). The Courts may make up an order between three and twelve months. Once a young person has been given a Referral Order they are referred to a Youth Offender Panel (YOP). The YOP meet outside the formal settings of the Court and they are made up of community volunteers and a trained representative from the local Youth Offender Team. The victim or a representative (for example, in the case of shoplifting a local business person may be present) are invited to attend the panel meeting. At the meeting the panel draws up a contract, which, when signed by all parties, activates the Order made by the court. The YOPs are guided by restorative justice principles of 'restoration, reintegration and responsibility' (Home Office, 1997).

For further reading on Referral Orders see:

Crawford, A. and Newburn, T. (2003) *Youth Offending and Restorative Justice: Implementing Reform in Youth Justice*. Cullompton: Willan Publishing.
Muncie, J. (2004) *Youth and Crime*, 2nd edn. London: Sage.
Newburn, T. (2003) *Crime and Criminal Justice Policy*, 2nd edn. Harlow: Pearson Education Limited (Chapter 8: Youth Crime and Youth Justice).
Smith R (2003) *Youth Justice, Ideas, Policy, Practice*. Cullompton: Willan Publishing.
www.youth-justice-board.gov.uk

Activity 9e: Drug Treatment and Testing Orders

Drug Treatment and Testing Orders (DTTOs) were introduced as part of the Crime and Disorder Act 1998, as a new community sentence available to the Courts for sentencing drug dependent offenders whose offences primarily related to their use of drugs. The Orders became available nationally in October 2000 and are a key component of the government's drug strategy, Tackling Drugs to Build a Better Britain. The Orders are available to those over the age of 16 and last for not less than six months and no more than three years. They involve attending a programme for at least 20 hours per week in the first 12 weeks of the Order reducing to less thereafter. The Order involves both drugs treatment and regular drug testing for the duration of the Order.

For further reading on DTTOs see:

Bean, P. (2002) *Drugs and Crime*. Cullompton: Willan Publishing (Chapter 5: The Drug Treatment and Testing Order and Drug Courts).
www.drugs.gov.uk

Activity 9f: Custodial sentences

To what extent can we justify the increased use of imprisonment as a form of crime prevention? What are the problems associated with use of custodial sentences?

Unlike the last two exercises, where we gave you particular Court Orders, in this exercise we would like you to consider the use of custodial sentences. At the time of writing, mid-2004, the prison population stands at over 75,000 and the Home Office estimates that this figure will increase to over 100,000 by 2009. As noted by Newburn, 'What has happened to the prison population in the past

10 years is, by any standards (except possibly those of the United States), extraordinary' (2003: 47). The increase in the prison population has been dramatic and according to Cavadino and Dignan, 'The English penal system is unjustly and irrationally harsh' (2002: 1).

KEY TIP

You may also want to pay particular attention to the **incarceration** of young people (Secure Training Centres now exist for young people aged 10–17).

Activity 9g

Using the following extract from Muncie and McLaughlin (2001: 15–16) *The Problem of Crime*, consider the ways in which the social construction of crime could be used to explain the various issues discussed in the Activities above. Note the key points that you would use to develop an argument; this provides the evidence to support an academic essay or examination answer. (Later look at Appendix 2 for a rationale for this activity.)

Crime as Social Construct

A vast array of behaviours have been (or can be) deemed 'deviant' or 'criminal' because they violate legal or normative prescriptions. But there is no common behavioural denominator that ties all of these acts together. Propositions, such as society is based upon a moral consensus or that the criminal law is merely a reflection of that consensus, also remain contentious. The interactionist school of sociology, for example, argues that there is no underlying or enduring consensus in society.

(Continued)

(Continued)

Rather the social order consists of a plurality of social groups each acting in accordance with its own interpretations of reality. Such diversity is as likely to produce conflict as much as consensus. Interpretations of reality are learnt through the ways in which people *perceive and react,* either positively or negatively, to the various behaviours of others (Mead, 1934). Thus, with respect to crime, an interactionist position would argue that defining crime with reference to legal or norm-violating actions is seriously limited. Rather crime is viewed as a consequence of social interaction: that is, as a result of a *negotiated process* that involves the rule-violator, the police, the courts, lawyers and the law-makers who define a person's behaviour as criminal. Behaviour may be *labelled* criminal, but it is not this behaviour in itself that constitutes crime. Rather behaviour is *criminalized* by a process of social perception and reaction as applied and interpreted by agents of the law. Crime exists only when the label and the law are successfully applied to an individual's behaviour. It is not what people do, but how they are perceived and evaluated by others, that constitutes crime. Whereas law-violation approaches argue that the existence of 'crime' depends on the prior existence of criminal law, interactionism logically contends that, without the *enforcement and enactment* of criminal law (or social reaction to certain behaviours), there would be no crime. As Ditton put it: 'the reaction is constitutive of the criminal (or deviant) act. In fact the reaction *is* the "commission of the act"' (Ditton, 1979, p. 20). Just as criminal law is constructed within society, then crime too is a social construction (Hester and Eglin, 1992). In this sense, society creates crime because it (or at least those in positions of authority) makes the rules, the infraction of which causes crime. As Becker explains, the social construct argument is not so much concerned with locating the causes of crime and deviance in social factors or social situations, but in establishing that:

Social groups create deviance by making the rules whose infraction constitutes deviance, and by applying those rules to particular people and

> labelling them as outsiders. From this point of view deviance is not a
> quality of the act a person commits, but rather a consequence of the
> application by others of rules and sanctions to an 'offender'. The deviant
> is one to whom that label has successfully been applied; deviant
> behaviour is behaviour that people so label. (Becker, 1963, p. 9)

Thus 'crime' has no universal or objective existence, but is relative to the subjective contingencies of social and historical circumstance. This in turn opens up and expands the range of criminological inquiry away from behavioural questions — Why did they do it? — towards *definitional* issues — Why is that rule there? Who created it? In whose interests? How is it enforced What are the consequences of this enforcement? (Cohen, 1973, p. 623). It implies that we will only come to understand why an action is regarded as criminal by examining both the processes of rule creation and law enforcement. We will not discover 'crime' simply by looking at behaviours, for there is nothing intrinsic to any behaviour that makes it criminal (Phillipson, 1971). We will not discover 'crime' simply by looking at violations of legal statutes. These constitute only one element of rule-breaking and in themselves are the subjects of social perceptions and interpretations. Similarly, asking the perennial question, 'what makes some people commit crime?', is short-sighted because there are no categories of activities that are inherently criminal (Sharrock, 1984, p. 99). In this way, the interactionist approach refutes the notion that criminality is driven by some particular motivation or that criminals are a species apart. Rather it contends that criminality is *ordinary, natural* and *widespread* and as a result requires no more explanation than that which might be attached to any 'ordinary' activity. However, what does require explanation is the complex process by which agencies of social control are able to construct a public identification of *certain* people as criminal, and how social reaction and labelling are able to produce and reproduce a recognizable criminal population.
A corollary of this, of course, is that the more such labels are applied and enforced, then the greater the chance that more 'crime' will be discovered.
In these ways, a social constructionist position continually alerts us to the 'constantly problematic, changing and contested nature of crime and social problems' (Young, 1999, p. 40).

(Continued)

(Continued)

References

Becker, H. (1963) *Outsiders: Studies in the Sociology of Deviance*. New York: Free Press.

Cohen, S. (1973) 'The failures of criminology', *The Listener*, 8 November.

Ditton, J. (1979) *Contrology: Beyond the New Criminology*. London: Macmillan.

Hester, S. and Eglin, P. (1992) *A Sociology of Crime*. London: Routledge.

Mead, G.H. (1934) *Crime Law and Social Science*. New York: Harcourt, Brace Jovanovich.

Phillipson, M. (1971) *Sociological Aspects of Crime and Delinquency*. London: Routledge and Kegan Paul.

Sharrock, W. (1984) 'The social realities of deviance', in R.J. Anderson and W. Sharrock (eds), *Applied Sociological Perspectives*. London: Allen and Unwin.

Young, J. (1999) *The Exclusive Society*. London: Sage.

9.3 Summary

- This chapter has introduced you to a range of theoretical perspectives.

- These theories have been linked to legislation and practical concerns about crime.

- You should be looking to make connections between the two elements.

- The key skill here is developing a critically analytical approach to 'criminological facts'.

Part Three

Looking Back and Moving On

In Part 3 we will summarise the key points that have been made in relation to studying Criminology at university. We will draw together some of the issues that have been discussed and focus on researching your work, presenting your work and taking assessments. Following the guidance and advice given here should help you to get the most out of your studies. There is also a discussion about ways in which you can make use of your degree.

Increasingly, a key feature of Higher Education is the importance placed on transferable skills by employers. There are a number of organisations seeking to encourage the development of these skills, for example, the University Vocational Awards Council (UVAC) and the Sector Skills Council for Criminal Justice. In this final section of the book we draw your attention to these vocational skills and look at how you can record and evidence this for future reference when applying for employment.

The two chapters in this section look at:

- **Reflections on Studying at University**

- **Making Your Degree Work for You.**

10 Reflections on Studying at University

The main aim of this book has been to take you on a journey through a Criminology degree programme in Higher Education.

We began by suggesting ways in which you could make the transition from school, college, work or family commitments. This was followed by guidance on how to study at university. The section on assessment provided a rationale for certain forms of assessment and provided the advice needed to produce work of a good standard. The importance of reading for a degree cannot be underestimated and in the chapters on finding information and essay writing we identified ways in which you could get the most out of your research.

We encouraged you to develop analytical skills in Chapter 9 by drawing your attention to the relationship between theoretical perspectives and policy and legislation relating to crime and criminal justice.

Of course, none of this advice will work without your contribution. We hope that the enthusiasm and excitement that you as students exhibit when first arriving as undergraduates will continue throughout your academic careers. By providing you with an understanding of what it means to study this discipline at university we hope that your choice of degree reflects your interest in crime and disorder, anti-social behaviour, criminal justice or other key debates in Criminology more accurately. This should enable you to be more focused and enthusiastic for the discipline. For those of you choosing to combine Criminology with other subjects, the focus on potential careers should have helped you to make appropriate combinations.

One of the most important skills that we have tried to encourage is that of **reflexivity**. Reflection requires you to think about what you are doing, why you are doing it and ultimately how well you are doing it. Throughout a degree programme this enables you to map your own progress. Remember, gaining a degree is not about competing with other students – it is about you gaining experience, knowledge and personal progress.

The skills that you can develop while you are at university can be relevant to specific careers when they are combined with subject specific knowledge, but, even if you choose a career that is not related to the degree you studied, many of the skills in relation to time management, self-directed study, research and critical analysis will be of benefit to you.

While we have focused on many skills that will help you to succeed in your studies we do not expect you to follow our guidance rigidly, but we do believe that those of you who use our suggestions and ideas as a template for your own degree programme will improve your performance.

By focusing your attention on how to gather data and make full use of the range of resources available, we have given you the opportunity to organise your personal study to suit your lifestyle, commitments and the requirements of your course. Contrary to popular belief, successful students are not always the ones who hibernate for three years in the University Library or Learning Resource Centre. It is the students who successfully balance the competing demands of contemporary society who frequently achieve their potential; indeed, in many cases you can exceed your own expectations if you get the balance right.

For this reason the chapters on assessment encourage you to think about how best to study, while presenting you with guidance on the expectations that universities have of you. By discussing learning outcomes and marking criteria, we have given you the opportunity to reflect on the work you produce and self-assess the quality of what you are producing.

From our own experiences as students and academics, we have identified some of the pitfalls that commonly trip students up such as:

- **being disorganised**

- **leaving everything to the last minute**

- **not answering the question**

- **not producing what is required**

- **plagiarising the work of others**

- **simply not engaging with the degree.**

There are, of course, many other issues that will present you with problems and this book has also provided some advice on these. As an undergraduate you need to be able to organise your life in such a way that allows sufficient time to study. Those of you who are mature students or students with family commitments often need to be much more organised to ensure that you can find time to study; after all, when your children need feeding or support with school work it is difficult to make excuses for not being around. Using the time when you are in university sensibly then is vital as this is often when you can get the most work done.

It is equally important to find time to relax and while we would not claim to be experts in relaxation techniques we have tried throughout this book to highlight the need to consider how and when to build in leisure time. This is as important as study time because often when you relax ideas that you have been struggling with frequently become clearer. This is why many academics have a note pad by the bed so that they can record those 'inspirational' thoughts when they wake with a start in the middle of the night!

In addition to discussing generic skills we have developed some criminological topics as this book has a two-pronged format. This provides you with an introduction to some of the key theories and concepts that you will study while at university. These have been presented in a straightforward and concise manner and to gain the knowledge to be successful you will need to read about these topics in greater depth. We have included some examples that are more complex and you may find these useful as your studies progress. The bibliographical information that is included is one of the most valuable resources available to you as this identifies texts commonly used on Criminology degrees. You can use this as a starting point for further reading and it may supplement or complement the reading lists you receive when you begin your degree. Later in this chapter we will provide some further reading to help you get started.

For most of you reading for a degree in Criminology will be the starting point of a career in Criminology related employment. Some of you may have a career in mind before beginning your studies while others will develop an interest or become expert in a specific area as a consequence of studying Criminology. Either way you need to present yourself effectively to employers if you are to get your feet onto the first rung of the employment ladder. Reflecting on your skills through the use of a personal development file is a good starting point. From this you can produce a curriculum vitae (CV) that will mark you out as a potential employee. If good, your CV is well presented, clear and concise and increases the chances of an interview; it is well worth the effort to produce it. Poorly constructed CVs are more likely to be overlooked and without an interview, no matter how good you are, you cannot persuade someone to employ you.

Used as supporting material to the work you do on your degree programme, this book will provide you with advice, guidance and the skills to make the most of your time at university. By adapting our activities and reflecting honestly on your own performance you should be able to balance all areas of your life to enable you to succeed at university and to enter the workforce in your chosen career with an opportunity to quickly progress. In the contemporary workplace 'life-long learning' is a key concept, and the study skills developed in Higher Education will ensure that you are able to engage with this as your career develops. This is particularly so with the contemporary emphasis on life long learning. There is an expectation that employees continually update their skills as their career progresses and it is quite likely that you could find yourself returning to university later in your career.

Perhaps the most important message we are trying to send through this book is that studying is a skill in itself. As such, it can be learned and practised and this enables you to develop good habits, to vary your experience and ultimately to engage in the experience of being an undergraduate studying Criminology, especially when the topic is so high on the agenda of politicians, the media and the public in general. You should not forget, however, that the study of Criminology for its own sake has merits and you do not need to see your degree programme as simply a vocational course of study.

Finally, we will identify further reading and, in a glossary, introduce some of the terms used in this book and in Criminology. As the glossary is not intended to be a dictionary on the topic or to provide a comprehensive list we have drawn mainly on some of the issues raised throughout the book; there are already very good books available to provide a more extensive source of information on criminological terms.

Making Your Degree Work for You

CHAPTER OVERVIEW

By the end of this chapter you should be familiar with:

➤ personal development files
➤ employability skills and employment opportunities
➤ preparing a curriculum vitae.

While many people still believe in the notion of 'knowledge for its own sake', degrees are increasingly seen to be a route to specific occupations. An increasing number of degrees have a vocational element to them and experience of work, the acquisition of skills for the workplace and occupational knowledge are the main things potential employers look for. Some academic disciplines are more suited to this than others and it is important that you look at ways of making sure that the degree you study for meets the requirements of the occupation you intend to pursue after graduation. Some aspects of this may be integral to your programme of study while others will need to be instigated by your own efforts. This chapter aims to look at some of the things that you can expect to be part of the degree programme and also offers some advice about gaining valuable practical experience alongside your studies. This can go some way towards addressing the problem of having vocationally related knowledge but no relevant work experience.

11.1   Personal Development Planning

The majority of students will have some experience of keeping a record of achievement or some form of reflective diary as these are regularly used in schools and colleges to assist in the presentation of your study skills, academic success, extra curricular activity and work experience. From September 2005, it will be a requirement of the Quality Assurance Authority (QAA) that all universities facilitate the process of Personal Development Planning (PDP) as part of the Higher Education Progress File. How this is introduced, maintained and developed will vary between institutions; however, it is important from your own point of view that you engage fully with this process. Over time, it is likely that potential employers will become familiar with these documents and will require you to present them alongside your curriculum vitae (CV).

You will be required to maintain a personal development file or progress file and, at whichever institution you attend, there will be a policy in place to implement this process. The file will contain two key elements:

- **A transcript of progress – this is a record of your learning and achievement and will include a record of your grades for all modules and your final degree classification. You may wish to include other certificates or qualifications gained in this section.**

- **A system by which you can reflect, monitor and build upon your personal development; for example, identifying activities carried out in a voluntary capacity and reflecting on the skills you feel you have developed.**

One of the main aims of this process is to link academic learning, personal achievement and practical experience so that graduates can demonstrate their suitability for employment. Increasingly, employers are identifying a range of key skills that they expect graduates to have gained. The fact that an individual has successfully graduated does not necessarily demonstrate the acquisition of these skills. Universities are increasingly attempting to develop their programmes of study to reflect National Occupational Standards. These are skills that are identified by employers groups at a national level and highlight the skills, knowledge and understanding required in employment and indicate the outcomes that demonstrate competence in specific occupations. Within the criminal justice field the Sector Skills Council for Criminal Justice and Skills for Justice (http://www.skillsforjustice.com) are working with employers and universities to develop these standards, and the personal development file will allow you the opportunity to provide evidence of your achievements and expertise in relevant areas. This is an attempt to break the cycle in which graduates have difficulty entering into careers because they lack experience and being unable to gain experience because they cannot gain practical skills in the workplace; in short, this is an attempt to ensure that graduates possess the employability skills required by employers in the twenty-first century.

KEEPING YOUR PERSONAL DEVELOPMENT FILE

How you keep and develop your Personal Development File will vary between institutions; some may use the personal tutorial system as a way of supporting this process, others may build it into the programme as a whole. Whichever way this is done it is likely that the responsibility for maintaining a file or portfolio will rest with you. It is important to note that in some universities this process may be incorporated into the formal assessment of your degree; that is, you will be assessed for some or all of the work that you present in your PDP. When you arrive at university you will have a range of skills and experiences that you have already developed and acquired; these will vary depending on your age and experience. However, during your studies you will develop these further by gaining new skills, knowledge and experience.

Activity 11a

As a starting point for your personal file make a list of your skills and experience when you start your degree programme:

- List your skills in the areas of communication (written and oral), numeracy and the use of Information Technology (IT).
- What experience have you got? For example: voluntary activity, paid employment, membership of organisations and leisure activities or hobbies.

To help you with Activity 11a here is some advice and some examples of things you may want to include.

Communication skills:

- *Written skills* You should consider whether you have written essays, reports and letters and if so were they as good as you would have liked, and how confident you feel in your writing skills. You should not be afraid to identify areas in which you consider yourself to be weak. Remember even if you are a student who left school or college some years ago you may have acquired some of these skills in the course of your work; these should be included here.

- *Oral skills*   This includes making presentations, speaking to an audience and contributing to discussions. Some of these skills are practised in schools and colleges. However you may have developed them in areas outside this – many jobs require you to communicate orally and this can be identified here. In addition, if you have been involved in amateur dramatics, for example, you may well have acquired some appropriate skills.

Remember that identifying your strengths and weaknesses is not an indication of success or failure but will help you to see where you need to concentrate your efforts to gain new skills or improve your existing ones.

Numeracy skills:

- Many people will have studied maths at school or college and this is a good indication of your ability in this subject; however, even for those who find maths difficult it is possible to provide evidence that you are numerate.

- You may be skilled at mental arithmetic even though you struggle with other aspects of maths. Often in a discipline like Criminology this may be regarded as an appropriate level of competence although when analysing crime statistics, for example, a greater level of numerical competence may be appropriate.

- It is important to remember that numeracy is a necessary skill in Criminology. Statistics are often used and an awareness of the relationship between figures enables criminologists to make judgements about levels of crime and the effectiveness of policies.

- Always remember that statistics can be used in many ways and your skill with numeracy helps you to be more analytical when using them.

IT skills:

- Many people beginning a degree programme have experience of using computers; this may have been at home, school and college or in the workplace.

- Remember that you will use a computer for more than word-processing while you are at university. Ask yourself if you are competent at sending emails, using the Internet, and accessing the electronic information systems used in libraries and learning resource centres.

- Remember that there are other technologies available and a range of applications that you may need to use; you may have used an overhead projector (OHP), or you may have produced and delivered a PowerPoint presentation.

- It is a fact that, while many of us would claim to be computer literate, we frequently fail to make full use of our computer's capacity. This is perhaps because the capabilities of modern machines are progressing more quickly than we can keep pace with.

From the starting point in Activity 11a you should have a basic assessment of your strengths and weaknesses in relation to a range of key skills and where you feel that your skills need to be developed. Identify ways in which this can be done. Some of this may be built into your degree programme in the form of study skills modules or research methods modules. Identifying your skills enables you to produce a CV and make job applications that match you to the person specification for a variety of jobs. Of course, employers are looking for more than skills; they want you to be able to demonstrate some relevant experience and this is often the case with occupations in the criminal justice system. You also need to include this in your personal development file.

Paid and voluntary work:

The easiest way to demonstrate that you have relevant experience is to show that you have previously worked in a similar role. For many students this is easier said than done, for several reasons. In some cases, you will have come to university straight from school or college; in others, you may have worked for a number of years but not in a related discipline. This does not mean that you do not have relevant employment skills and there are opportunities for you to gain experience while you study. In some universities this will be more formal than in others, but you can gain work experience in a variety of ways. Before we discuss how to gain experience look at Activity 11b to produce a self-assessment of your employment skills.

Activity 11b

Look at the following questions and construct a list of your experiences:

- Have you ever been in paid employment? If 'yes' list the jobs you have done and give a brief outline of what this type of work involved.
- Have you ever undertaken voluntary work? List the places you have done this in.
- What leisure activities do you engage in? Make a list of groups, clubs and sports teams that you have been involved with.

If you have answered 'yes' to the first question and have a list of jobs, think about the skills you need or needed to do this type of work. Even if you were doing part-time work while you were at school or college you will have developed some skills that will be of use in the future. For example working in a fast food outlet or a bar will have required you to develop some skills in relating to people. Here, also, money handling is evidence of your honesty and working as part of a team can show your reliability in a work situation. You need to identify such skills and link them to the type of work you are looking to do in the future.

Another way in which you can develop employability skills is through voluntary work. Sometimes this kind of activity is formalised; you may be a registered volunteer with a local or national charity and this can be used as evidence of skills in the workplace. Another way of becoming involved in voluntary work could be helping people within your own community, such as residents groups or community groups, or through activities sponsored by your school or college. This type of activity may well be something you can continue when you get to university. Many universities support voluntary activity in a formal way but you can make a contribution to the community in which you live that will be useful in developing your skills. Some examples of this with specific reference to Criminology could include:

- **working on youth offender panels — you will be able to get information about this from your local Youth Offender Service**

- **voluntary work with the probation service — you can find out about this by contacting the National Probation Service in your area**

- **custody visiting with the local police — your local police station will be able to tell you where to get more information**

- **working with the local Youth Offender Service — contact the one in your local area for more information.**

Depending on the area in which you study, these opportunities will be available to you and you will be able to find out more from the local press, the university and perhaps your student union. Many universities now have good links with local organisations looking for people to undertake voluntary work. The Council for Voluntary Service (http://www.nacvs.org.uk) in your area can also put you in touch with some of these groups and organisations.

By now you will be getting the message that the experience and evidence that you can provide to support an application for a job can come from a variety of sources. Most of you are or have been engaged in leisure activities, such as sports or belong to clubs, and often the things we do here can provide evidence of certain attributes that employers are looking for. Here are some examples in a generalised format; individually you may well be able to develop this further.

LEISURE ACTIVITIES

Most students have been involved in some form of leisure activity and whether this was in a group situation or as an individual there will be skills and personal attributes that can be identified to support employment applications. The wide range of activities that we might identify as leisure is too vast to include in this discussion but we can present some examples. It is worth noting, of course, that leisure to one person may seem like work to another so we need to think about what we gain from our activities rather than how other people may classify them. For the purposes of this chapter leisure will include sport, recreation, pastimes and hobbies; however, there are other activities you may wish to consider in relation to your personal development.

Sport can be seen as recreational in several contexts. You can play sport, coach sport, watch sport or be a sports administrator and whichever of these activities you engage in employers will be interested in the skills, lessons and attributes that demonstrate an ability that may be of value them.

Employability skills are not simply the skills gained in paid employment, they are also the transferable skills you can acquire in many areas of your life.

From the point of view of playing sport there are a number of examples:

- Team sports develop teamwork, communication skills and an ability to work co-operatively with others. Captaining a team demonstrates leadership skills, especially if the team is successful!

- Individual sports can demonstrate an ability to work on your own initiative, a determination to succeed as the training and practice is often harder, and a capacity to be focused.

- Coaching sports requires communication skills, an ability to develop tactics and an interest in supporting other people to be successful.

- Many sports clubs have committees and serving on one of these can demonstrate administrative and organisational skills.

Of course many of us are not interested in sport and prefer to engage in other forms of leisure pursuits. In some instances, this involves membership of clubs, societies and organisations and there is the potential for demonstrating some of the skills mentioned above.

- Amateur dramatics and opera groups require teamwork, practice and direction and communication.

- Photography clubs, reading groups and other specific interest groups usually have some form of committee to organise activities, advertising and peer support. The development of skills is often through sharing information and showing others how to do things.

- For some people their leisure is based around an individual activity; perhaps learning a musical instrument, developing skills in a practical activity or playing computer games. There are skills involved in this and the determination and commitment to practise can often be something that would be seen positively by employers.

You should by now be able to see that there are many ways in which we develop skills, not all of them in formal learning situations as in some cases it is simply by doing something we enjoy as a hobby. When you begin to think about compiling a personal development file you need to think much more broadly than perhaps we normally do when we think about the skills employers are looking for. The following section of this chapter will look at some of the occupations that criminologists may be attracted to and will discuss some of the skills that may be needed. This will not be an exhaustive list or a complete inventory of skills but should enable you to think about what to do when you graduate and how to develop the skills you may need to be successful.

11.2   Criminological occupations

For many people the discipline of criminology is a direct route into the various agencies of the criminal justice system:

- **the Police**

- **the Courts**

- **the Crown Prosecution Service**

- **the Probation Service**

- **the Prison Service.**

These are perhaps the most obvious examples although there are many other organisations aligned to these. Some examples would be:

- **drug referral workers, often based in police stations**

- **victim support organisations**

- **witness support schemes**

- **domestic violence support groups**

- **The Youth Offender Service.**

As the criminal justice system develops, the range of occupations is also expanding. In recent years through Crime and Disorder legislation Local Authorities, the Local Police Service, the Local Health Authority and the Fire Service have been required to develop Crime and Disorder Partnerships; this opens up a range of opportunities for criminologists to become involved in projects, crime reduction initiatives and specialised work with offenders and potential offenders. During 2003 and 2004 there has been an increased awareness of the problems associated with anti-social behaviour and there has been an increase in employment opportunities associated with attempts to reduce the incidence of such behaviour. Alongside these developments is the need to evaluate the effectiveness of these initiatives, commonly referred to by government departments as 'what works' and this evaluation is based on research carried out by criminologists. Of course, many of you will not know what careers you might be interested in either when you start your degree or when you graduate. The skills you acquire when studying will be helpful at a later date when you are more sure of the direction you intend to take.

Activity 11c: Criminology and occupations

- Making use of newspaper advertisements, Internet job-search sites and recruitment information make a list of potential occupations that your criminology degree will be useful in. (There is no right answer and inevitably you will make some decisions based on your personal interests.)
- Take one or two examples from the list; these could be areas in which you are interested but do not have to be. Now list the personal qualities, skills and knowledge that are seen as important for potential employees.
- Match these to the list of personal experience and skills that you have identified earlier. (Activities 11a and 11b)
- You should be able to produce a statement of your own experience that matches those required for the job. Identifying where you still need experience or training can be the basis for your studies and any voluntary work you do in your degree.

11.3   Continuing in education

Increasingly, the workplace requires individuals to continue training throughout their careers and in many cases this involves short courses or professional qualifications related to the occupation you enter. The National Probation Service is a good example of this; graduates who join the National Probation Service are required to undertake further education at university. Often those who have studied Criminology can be exempted from some modules as they have already covered this work. Social work and the caring professions are other examples of this. Those seeking careers in the legal profession will need to undertake further study before being qualified to practise Law.

Other graduates may well feel that they want to continue in Higher Education for other reasons. They may want to undertake further study in the criminological field as a postgraduate student or to undertake a research degree. There is a range of programmes available at this level and for many the first step is a Masters degree. Master of Science (MSc) and Master of Arts (MA) are usually taught degrees in a specific subject area, for example, Criminology, Criminal Justice, Policing or Research Methods. Usually for one year (full-time) or two years (part-time) these programmes allow you to develop greater expertise in your chosen field. Research degrees, usually Master of Philosophy (MPhil) or Doctor of Philosophy (PhD), are based around an extended research project and last for three years or more depending on whether they are full- or part-time.

MATCHING SKILLS TO OCCUPATIONS

Look at Activity 11d. It is not the intention here to exactly match skills to specific occupations but to get you thinking about how the skills you have identified earlier can match to job descriptions relating to some of the jobs in the field of criminal justice. In this sense this is a qualitative exercise rather than a quantitative one. In other words, you are not simply ticking boxes but are giving some thought to how your own skills will enable you to further develop your career.

Activity 11d

Identify two or three careers that you feel would be interesting —
perhaps even look for job advertisements and send for the application pack.

There will usually be a person specification that identifies essential and
desirable qualities for the successful candidate. Attempt to match your own skills
and qualifications with this. This will help you to identify areas in which you need
to gain experience or further qualifications. You can do this at any time during
your degree to assist in planning.

11.4   Constructing a curriculum vitae (CV)

This is an important part of the process in gaining employment when you graduate.
In most cases the first contact you have with a potential employer is based on your
CV; this is your first opportunity to make a good impression. This is when you can
'sell yourself' to a potential employer. Concentrate on your strengths and make sure
that your CV is:

• **pleasant to look at**

• **easy to read.**

Potential employers will be looking at dozens or perhaps hundreds of CVs when they
are shortlisting for an advertised vacancy and may receive many more unsolicited
enquiries. In the main your CV will only be scanned and it is important that the
reader is able to identify the key points they are looking for. While much of your CV
will be based on factual information it provides the opportunity for you to give some
clues as to the sort of person you are. This may help an employer decide how well you
will fit in with the existing workforce.

KEY TIP

Try to make your CV interesting and lively.

As each CV relates to an individual it is not possible to identify a definitive style. The layout and content are for you to decide and may well be dictated to some extent by the position for which you are applying. However, the information you have gathered in your Personal Development File will be of use here as you can extract the important information and summarise it in your CV. Having said that there is no one style of CV, there are some general rules and some information that is essential.

Paper colour should be white, cream or pale blue – bright colours may attract the reader's attention but are more likely to be discarded than read. The length of your CV should be no more than two pages. Providing that you have included all relevant information, you could produce a one-page document. While the use of graphics may make your CV stand out it is the words that count; do not include anything that is irrelevant to the post for which you are applying. Present the information in short, easy-to-read blocks that are clearly headed and distinguishable from each other. The information should be presented in a logical order making it easy to follow. Do not leave blank spaces apart from in between sections. Wherever possible you should use positive language as this can make the difference between an ordinary CV, a good CV and an excellent CV. Use verbs that suggest strengths such as influence, supervise, problem-solve, time-manage and delegate. While there is no set format for a CV the following sections need to be included.

PERSONAL DETAILS

Generally speaking this section writes itself: you must include your name, address, telephone (home and mobile) and email address.

KEY TIP

Don't use a 'silly' name for your email address; it is unlikely to be seen as professional and in some cases may cause offence.

Your date of birth, marital status, nationality and your gender are not necessary unless they are relevant to the specific application.

PERSONAL STATEMENT/CAREER OBJECTIVES

You can use this as an opportunity to highlight particular skills, previous experience or your career aims. This is not a lengthy discussion of your past and future but a succinct (three or four sentences) summary of what you can offer the employer. Some of the points you might want to raise here could be the basis of your covering letter, which we discuss later in this chapter.

EDUCATION

Whether you are applying for employment appropriate to a graduate or not, a potential employer will be interested in what you have been doing most recently. You should list your education and dates beginning with the most recent, then college/school. It is not necessary to go further back than schools since you were eleven; for mature students even this may not be necessary. You should provide some discussion of the nature of your most recent study and write this from the point of view of what you got out of it rather than copying from the prospectus.

EMPLOYMENT

This is the most important part of your CV as this is where you get the opportunity to demonstrate your ability to function in the workplace. You need to highlight the employment experience that you have specifically related to the job for which you are applying as well as identifying other experience that you may have. If you have a lot of work experience be selective about the jobs you highlight. Again, a chronological list is useful but do not go too far back or include too many irrelevant posts. You can highlight specific experiences that the employer may wish to pursue in an interview. Mature students sometimes need to demonstrate a solid history of employment before study. This does not usually go back more than ten years unless there is a position relevant to the post you are interested in.

For those who have no formal employment history due to being carers or parents you should identify roles played in school activities, membership of clubs or other voluntary activities.

The above are essential in any CV but there are some areas that you may wish to highlight to strengthen your application. Do not mention them if they do not apply but where they can add weight to your application they should be included. IT skills and language skills will be an advantage to you; as will any positions of responsibility, some examples of which are given below.

Positions of responsibility

- captain of a sports team

- course representative

- young enterprise experience

- a position held in a club/society

- sports coaching

- youth leaders

- member of Youth Offender Panel.

Appendix 1: Sample Answers to Activity 7a

Activity 7a asks you to identify ten key issues/debates from the synopsis on youth crime. The following is an example of ten such areas. It is important to note, however, that these are *not* the only ten issues/debates that can be identified; you may well be able to think of others:

1 Single parent families   What does this term mean? Is the relationship between crime and one-parent families strong? If so, is it consistent across all single parent family types?

2 Why do young people commit crime?   What does research show in relation to why young people commit crime? Boredom? Buzz? Drugs?

3 Drugs and crime   (In your answer you may choose to focus on one of the possible reasons discussed in 2 above, either because the evidence is strong or is a recurring theme within the literature.) What is the role of drugs? What is the relationship between drug use and crime? (You need to keep this brief, however, and always relate it to the question. Therefore, one could argue, for example, that it is more about peer group, drugs and external factors outside the family that are more likely to play a role in shaping criminal careers.)

4 Underclass/moral decline   Is there an underclass? Has there been moral decline typified by single parent families? Is there an ideal family type?

5 Poverty and crime   Is it more to do with poverty rather than a decline in morality? Is the relationship between poverty and crime?

6 Crime types   What type of crime are we discussing? Is it that certain types of crime are linked? — visible street crime rather than white collar crime, for example.

7 Poverty, crime type and policing   It may be that because of poverty the young person lives in an area that is over-policed and they commit visible crimes and therefore are more likely to be involved in behaviour that is detected.

8 Responsibility   Who should be responsible for the behaviour — the child, their parent/guardian? What is the role of schooling and peer group?

9 External family factors   What is the role of the local school, employment market, youth service provision, for example? Are these issues more important than the family?

10 Discipline   What is the role of discipline? Is crime created because of families without fathers and therefore a disciplinarian?

Appendix 2: Rationale for Activity 9g

Activity 9g is asking you to consider the concept that crime is a consequence of social factors and the influence of social institutions. This is significant when we try to understand why, how and when criminal activity takes place. Essentially this challenges the notion that we can identify criminals using scientific research. Within this extract (which should represent a small part of your reading) we can find references and quotations that would be useful in constructing a debate around the key debates discussed in Chapter 9.

There is for example, Mead's (1934) concept of crime being a consequence of the way behaviour is perceived by others or indeed self-perception. As the authors suggest crime is a consequence of 'social interaction'. This challenges the view that criminals behave as they do because of genetic, psychological or physical factors, but rather as a consequence of their relationship to the communities in which they live.

Crime then is a not a universal phenomenon, concepts of what is and is not criminal vary across, time, place and societies and, to this extent, crime is a socially created phenomenon, rather than an absolute form of unacceptable behaviour.

Appendix 3: A Sample Degree Programme Structure

Here we present an outline of a degree programme over three years based on you studying six 20-credit modules

BSC (Hons) Criminology

Level 1			
Crime, Justice & Society 40 Credits	Sociology 40 Credits	Youth Studies 20 Credits	Images of Deviance/Study Skills 20 Credits
Level 2			
Criminal Justice 40 Credits	Criminological Theory 40 Credits	Planning & Designing Criminological Research 20 Credits	Option 20 Credits
Level 3			
Dissertation 40 Credits	Penality 20 Credits	3 Options @ 20 Credits	

OPTION MODULES INCLUDE:

Drugs and Society
Rape and Sexual Assault
Community, Crime and Crime Prevention
Policing and Social Control
Race, Crime and Criminal Justice
Youth, Crime and Justice
Interpersonal Violence
Victims of Crime
Domestic Violence
Psychology and Crime
Working with Offenders
Philosophies of Punishment
Crime, Violence and Masculinities

While this is the structure that is used at the University of Teesside other institutions will have a similar structure with differently named modules.

Appendix 4: A Sample Module Guide

This is a copy of a module guide. Most universities will provide you with this type of information and it is through this that you can identify what is expected of you throughout the academic year. The format may vary between various Institutions but the general content will always be similar.

Police and Society

The role of the police in maintaining order in society has been an area of debate for nearly two hundred years. This module examines the relationship between the police and the communities within which they work. However consideration is also given to the way in which a 'mixed economy' of policing in society has developed to include private security, community policing, community safety and vigilante groups. The module considers a number of issues in relation to the policing of society beginning with a review of the historical developments of the police force and the political and economic considerations that have influenced the nature and development of contemporary policing practices. The challenges facing the police as an organization are discussed and include issues such as accountability, autonomy, and police powers. Contemporary debates surrounding the issues of race, class and gender will also be included. The aim of this module is to enable students to develop an understanding of the functions of the police and the demands placed on them by an ever changing society in which rising levels of crime have contributed to an increased fear of victimization. Increasingly the nature of policing in Britain is influenced by global factors, threats from terrorist organizations and advances in technology. The local 'bobby' walking the beat is no longer seen as the way in which the police can be most effective, despite the wishes of society, and this module therefore explores some of the alternatives to traditional policing practices. Also there will be discussion of the way in which the role of the police in the maintenance of social order may be developed in the future. With these key issues as the focus students will be provided with material that enables them to develop an understanding of current developments in policy and practice.

Main Learning Outcomes

On successful completion of this module the student will be able to:

- Understand the dimensions of the Police function in society

- Recognise the impact of social, political, organisational and cultural factors on Police practice

- Evaluate the use of discretion and assess the dependence/independence of the Police in relation to the state and local communities

- Understand current developments in Police policy and practice

- Have developed skills in communication and critical analysis

Teaching and Learning

The module runs over one semester with a one-hour lecture and a one hour seminar each week. Lectures provide the substantive, conceptual and theoretical basics of the course and relevant issues are raised. Weekly seminars will allow students to explore key questions in greater depth. Reading will be required for seminars and relevant sources will be identified throughout the module. When appropriate students may be asked to work in groups or individually on presentations around specific topics. These are intended to allow the opportunity to develop more in depth discussion of the key issues/debates that are taking place and to contribute to the development of essay outlines for the module assessment. Student should ensure that they read relevant material in preparation for seminars as this aids informed discussion. A lecture outline and references to relevant information will be provided on the student Intranet site, full lecture notes will not be available electronically although any overheads will be available.

Assessment

Assessment consists of one 3000 word essay to be submitted on the 1st January 2007. Essay titles are included in this guide.

Marking Criteria (Level Two, Year Two)

70%–100%
Points are made clearly and concisely, always substantiated by appropriate use of source material. There is evidence of a sound ability to interrelate critically, theories with examples from practice. The work contains coherent arguments with some evidence of original thought. Presentation is excellent.
60%–69%
Very good presentation with an emerging ability to apply knowledge critically to practice. Appropriate evidence, good use of source material, which supports most points clearly. Content is wholly relevant, within a fluent coherent structure. Critical reflection could be developed further.
50%–59%
There is demonstration of a sound knowledge base, but limited critical and practical application of concepts and ideas. Content is largely relevant although points may not always be clear, and structure may lack coherence. Use of source material to illustrate points is generally adequate but may be lacking in some instances. Contains some critical reflection. The presentation is of a good standard, but with minor errors in grammar and spelling.
40%–49%
Adequate presentation with some errors. The work is descriptive but relevant, with clear evidence of knowledge and understanding. There is evidence of some reading and there is limited critical reflection. Links to practice are made, although arguments are often lacking in coherence and may be unsubstantiated by relevant source material.
39% and below – fail
Poorly structured, incoherent and wholly descriptive work. Limited evidence of appropriate reading, and no evidence of critical thought. Referencing poor or missing.

Attendance

University policy requires all students to maintain an 80% attendance level on this module. Records of attendance will be maintained and students will be contacted if they are absent for two sessions during the semester. If your attendance falls below 80% you may be deemed not to be taking the module, unless you are given specific exemption from attendance by the course leader, for which medical or supporting evidence will be required.

Lecture Programme

LECTURE ONE:

Introduction: This session will outline the key themes and issues for discussion on the module. An overview of the structure, organization, resources and content of the module will be provided.

Seminar topic: (week 2) Introduction to module and general overview.

The remainder of the lecture programme will be delivered in four blocks covering the following topics:-

BLOCK ONE

History, origins and development of policing and the police service.

Revisionist and traditional history of the police

Changing roles/power/discretion

Abuses of power and miscarriages of justice

Who are the police issues of equality/inequality?

LECTURE TWO:

The history and the origins of policing and the Police.

Seminar topic: (week 3)

Critically discuss the merits of the various perspectives of the development and history of the modern police.

Reading:

Johnston, L. (2000) Policing Britain: Risk, Security and Governance (Chapter 1)

Rawlings, P. (2002) Policing a Short History

Reiner, R. (1985, 1992, 2000) The Politics of the Police (Chapter 1)

LECTURE THREE:

The structure and organization of the modern police force.

Seminar topic: (week 4)

To what extent can it be argued that local police forces are unable to provide a local service when accountability is increasingly centralized?

Reading:

Johnston, L. (2000) Policing Britain: Risk, Security and Governance (Chapter 5)

Rawlings, P. (2002) Policing a Short History

Reiner, R. (1985, 1992, 2000) The Politics of the Police (Chapter 2)

LECTURE FOUR:

Internal and external accountability and abuses of power.

Seminar topic: (week 5)

Discuss the extent to which accusations of institutionalized inequality can be justified.

Reading:

Bowling, B. (1999) Violent Racism, Victimisation, Policing and Social Context

Bowling, B. and Phillips, C. (2002) Racism, Crime and Justice

Chan, J.B.L. (1997) Changing Police Culture: Policing in a Multi-Cultural Society

McPherson, W. Sir The Stephen Lawrence Inquiry

Reiner, R. (1985,1992, 2000) the Politics of the Police (Chapters 3, 4, 5)

Westmarland, L. Gender and Policing: Sex, Power and Police Culture

Young, M. (1991) An Inside Job: Policing and Police Culture in Britain

Politics, development and future of policing

Policing and the state

Private and public policing

Arming the police/police as a military force

Policing protest, industrial disputes and environmental issues

LECTURE FIVE:

The politics of law and order.

Seminar topic: (week 6)

Discuss he way in which Law and Order has become a significant issue for Politicians since the late 1970's

Reading:

Brake, M. and Hale, C. (19) Public Order and Private Lives: The Politics of Law and Order

Gest, T. (2001) Crime and Politics: Big Governments Erratic Campaign for Law and Order

Norton, P. Ed. (1984) Law and Order and British Politics

Reiner, R. (1985, 1992, 2000) The Politics of the Police

LECTURE SIX:

The development of private and public policing.

Seminar topic: (week 7)

To what extent has the balance between private and public policing changed in recent years?

Reading:

Button, M. (2002) Private Policing

Johnston, L. (1992) The Rebirth of Private Policing

Johnston, L. (2000) Policing Britain: Risk security and Governance (Chapters 5–9)

Jones, T. and Newburn, T. (1998) Private Security and Public Policing

Shearing, C.D. and Stenning, C. eds. (1987) Private Policing

LECTURE SEVEN:

Conflict of interests? Policing protest.

Seminar topic: (week 8)

Discuss the way in which the police's relationship with the public has changed as a consequence of the ways in which the Police have dealt with public disorder and protest over the last 30 years.

Reading:

Coulter, J. (1984) A State of Siege: Politics and Policing of the Coalfields

Fine, B. and Millar, R. eds. (1985) Policing the Miner's Strike

Green, P. (1990) The Enemy Without: Policing and Class Consciousness in the Miner's Strike

Johnston, L. (2000) Policing Britain: Risk Security and Governance (Chapter 6)

McCabe, S. (1988) The Police, Public Order and Civil Liberties: The Legacy of the Miners Strike

O'Byrne, M. (2001) Changing Policing: Revolution Not Evolution

International perspectives on policing

Globalization

Policing cyberspace

Asylum, immigration and the movement of labour

LECTURE EIGHT:

Globalization and transnational co-operation.

Seminar topic: (week 9)

Discuss the extent to which the increase in cross border crime, the global economy and international conflict have influenced the development of transnational policing.

Reading:

Anderson, M. et al. (1994) Policing Across National Boundaries

Anderson, M. et al. (1995) Policing the European Union

Sheptycki, J. (2000) Issues in Transnational Policing

LECTURE NINE:

The Internet, **cyber crime** and the role of the police.

Seminar topic: (week 10)

Is the police service in the 21st century equipped to deal with the increased opportunity for crime that has come about due to new technologies?

Reading:

Cullompton, Y. (2002) Dot Cons: Crime Deviance and Identity on the Internet

Wall, D.S. Ed. (2001) Crime and The Internet

BLOCK FOUR

The future of policing

LECTURE TEN:

Policing and Multi-Disciplinary partnerships

Seminar topic: (week 11)

What are the barriers to successful multi-agency partnerships?

Reading:

Matthews, R. (2000) Kerb Crawling, Prostitution and Multi-Agency Policing

Sutton, M. (1996) Implementing Crime Prevention Schemes in a Multi-Agency Setting

Taket, A.R. et al. (2000) Partnership and Participation: Decision Making in the Multi-Agency Setting

LECTURE ELEVEN:

Contemporary debates surrounding the future of the police.

Seminar Topic: (week 12)

Assessment preparation.

Module overview and final assessment preparation.

Seminar Programme

The aim of the seminars is to build on the substantive and theoretical input from the lectures. Students will be required to identify topical issues relating to aspects of policing around the major themes of the module. Students will be required to undertake reading and to prepare discussion papers for use in seminar sessions. As policy relating to the criminal justice system is constantly changing and the subject of constant debate topics may be introduced as they arise. In addition to group seminars students can arrange individual tutorials through the diary system.

Indicative Reading

The reading identified for seminar preparation is not a prescriptive list and provides only the basis for discussion, additional reading will broaden the student's understanding and will enable students to fully engage with this module. The following is an indicative list of relevant literature but this is not exclusive. Other texts will be equally as useful!

Alderson, J. (1979) *Policing Freedom*. Estover: McDonald and Evans

Bayley, D. (1994) *Policing for the Future*. Oxford: Oxford University Press

Bowling, B. (1998) *Violent Racism: Victimisation, policing and context*. Oxford: Clarendon Press

Bowling, B. & Foster, J. (2002) 'Policing and the police', In Maguire et al. (2002) *Oxford Handbook of Criminology*. Oxford: Oxford University Press. pp. 980–1033

Bowling, B. and Phillips, C. (2002) *Racism: Crime and justice*. Harlow: Longman

Brogden, M., Jefferson, T. and Walklate, S. (1988) *Introducing Policework*. London: Unwin

Heidenson, F. (1992) *Women in Control: The role of women in law enforcement*. Oxford: Oxford University Press

Jefferson, T. and Grimshaw, R. (1984) *Controlling the Constable*. London: Frederick Muller/Cobden Trust

Johnston, L. (2000) *Policing Britain: Risk security and governance*. Harlow: Longman Criminology Series

Jones, S. (1987) *Policewomen and Equality*. London: MacMillan

Manning, P. (1977) *Police Work*. Cambridge: Mass

McPherson, W. (Sir) (1999) *Inquiry into the Matters Arising From the Death of Stephen Lawrence*. London: Stationery Office.

Riener, R. (1992) *The Politics of the Police*. Sussex: Harvester Wheatsheaf.

Stephens, M. and Becker, S. (eds) (1994) *Police Force Police Service*. London: MacMillan

Wadddington, P.A. (1991) *The Strong Arm of the Law*. Oxford: Oxford University Press

NB. This list is not exclusive other texts are equally as relevant.

SUPPORTING MATERIALS/RESOURCES

The British Journal of Criminology

Social Problems

Feminist Review

Crime Law and Social Change

Howard Journal

Police Review

In addition there are Government web pages that provide information about police practice, policy and research. Links are available on the module Blackboard site.

Assessment Criteria

The assessment criteria are produced earlier in this guide and further information will be provided later in the module to support you in the preparation of your essays. However it is important to note the following general comments. The grades you are awarded reflect the quality of several aspects of an academic essay:

- **The standard of written English, use of grammar and spelling;**

- **The use of appropriate references and the use of a recognised referencing system, this is conventionally the HARVARD system of referencing;**

- **The demonstration of your understanding of the subject/topic area;**

- **The use of theoretical concepts;**

- **The use of evidence to support your arguments;**

- **Evidence that your work is based on more than just lecture notes.**

It is important to note that a well written, well presented argument that is not referenced or supported by evidence will not meet the criteria for the higher grades. You should be seeking to produce work that is of a high academic standard, rigorously researched and conforming to acceptable academic standards. You will receive support from your tutors in this area and should make use of the opportunity of individual tutorial support if you are in doubt about what is required of you.

NOTES ON REFERENCING:

Students are reminded that all written assignments must be fully referenced with a full alphabetical bibliography provided. The customary way of doing this is to use the Harvard **System of Referencing.** Failure to provide references of your sources can raise questions about plagiarism, which, if they are confirmed, may lead to penalties being imposed. Students should make themselves aware of the University's policy on plagiarism. This is something that will also be addressed in the seminar sessions.

EXTENSIONS AND MITIGATION:

Students are reminded that they must hand in assignments on the due date, any work handed in after the deadline may be subject to penalties. If there are problems about meeting the deadline an extension may be requested. However this must be before the due date (usually one week) and must be done on the appropriate form. When work is handed in late because of health problems a Doctor's letter or medical certificate is required, this also applies in the case of non-submission of work. It must be stressed that extensions will only be granted in cases of genuine difficulty. This does not normally include pressure due to several deadlines being close together or because books were not available in the library, **it is your responsibility to plan and manage your workload to ensure deadlines are met!!** When work cannot be presented due to health, personal or other extenuating circumstances students can request that their assessment is deferred. **Documentary evidence in support of mitigation must accompany all applications.** Where a deferment is granted the assessment will take place at the next opportunity usually the re-sit period in August.

Plagiarism

Your attention is drawn to the University policy on plagiarism, which is viewed as a serious matter. You should note the current policy, which can be found in the School of Social Sciences and Law student handbook.

Appendix 5: An Example of an Essay Mark Sheet Pro Forma

This is an example of a feedback sheet; this will vary between institutions but gives an indication of key areas to pay attention to:

ESSAY MARK SHEET

NAME   MODULE TITLE

	NOT PRESENT	WEAK	FAIR	GOOD	VERY GOOD
WRITING SKILLS/STYLE					
STRUCTURE					
STYLE					
REFERENCING					
BIBLIOGRAPHY					
GRAMMAR/SPELLING					
USE OF QUOTATIONS					
PRESENTATION					
CONTENT					
THEORY					
EVIDENCE					
CRITICAL ANALYSIS					
RELEVANCE TO QUESTION					
ARGUMENT SENSE					
RESEARCH					
BREADTH OF READING					
USE OF JOURNALS/ ELECTRONIC SOURCES					
CONTEMPORARY MATERIAL					

OVERALL COMMENTS

TUTOR   GRADE

Appendix 6: Useful Websites

The authors do not take any responsibility for the content of the websites.

Corporate and White Collar Crime

Corporate Accountability
http://www.corporateaccountability.org

Crime Prevention/Reduction

Crime Concern
http://www.crimeconcern.org.uk

Home Office Crime Reduction Website
http://www.crimereduction.gov.uk

International Centre for Crime Prevention
http://www.crime-prevention-intl.org

NACRO
http://www.nacro.org.uk

Courts and Criminal Justice

Audit Commission
http://www.audit-commission.gov.uk

Court Service
http://www.courtservice.gov.uk

Criminal Court Review
http://www.criminal-courts-review.org.uk

Criminal Justice System for England and Wales Online
http://www.cjsonline.org

European Court of Justice
http://europa.eu.int/cj/en/index.htm

Inspection of Court Services
http://www.mcsi.gov.uk

Magistrates Association
http://www.magistrates-association.org.uk

Domestic Violence

Black Women's Rape Action Project
http://www.womenagainstrape.net

Campaign Against Domestic Violence
http://www.cadv.org.uk

CPS — Domestic Violence
http://www.cps.gov.uk/legal/section3/chapter_c.html

Home Office Domestic Violence Pages
http://www.homeoffice.gov.uk/crime/domesticviolence/index.html

Men's Aid
http://www.crisisline.co.uk/mensaid

Women's Aid
http://www.womensaid.org.uk

Drugs

Alcohol Concern
http://www.alcoholconcern.org.uk

Drugscope
http://www.drugscope.org.uk

European Monitoring Centre for Drugs and Drug Addiction
http://www.emcdda.eu.int

Home Office Drugs Website
http://www.drugs.gov.uk

Institute for Alcohol Studies
http://www.ias.org.uk

Scottish Drugs Forum
http://www.sdf.org.uk

The Centre for Drug Misuse Research
http://www.gla.ac.uk/Inter/DrugMisuse

Ethnicity

Black Britain
http://www.blackbritain.co.uk

Commission for Racial Equality
http://www.cre.gov.uk

Criminology in the Millennium
http://www.ruthchigwada-bailey.inuk.com

Institute for Race Relations
http://www.irr.org.uk

Gender

Fawcett Society
http://www.fawcettsociety.org.uk

Women in Prison
http://www.womeninprison.org.uk

Homelessness

Crisis
http://www.crisis.org.uk

Human and Civil Rights

Amnesty International
http://www.amnesty.org.uk

British Institute of Human Rights
http://bihr.org

Civil Rights
http://www.civilrights.org

Inquest
http://www.inquest.org.uk

Innocent Until Proven Guilty (remand prisoners)
http://www.innocentuntilprovenguilty.com

Justice
http://www.justice.org.uk

Liberty
http://www.liberty-human-rights.org.uk

Miscarriages of Justice
http://www.mojuk.org.uk

Office of the United Nations Higher Commissioner for Human Rights
http://www.ohchr.org/english

Statewatch
http://www.statewatch.org

United Nations Children's Fund
http://www.unicef.org

Police

Association of Chief Police Officers
http://www.acpo.police.uk

Association of Chief Police Officers — Scotland
http://www.scottish.police.uk

Association of Police Authorities
http://www.apa.police.uk

Constabulary
http://www.constabulary.com

Criminal Records Bureau
http://www.crb.gov.uk

Europol
http://www.europol.eu.int

Gay Police Association
http://www.gay.police.uk

Her Majesty's Inspectorate of Constabulary
http://www.homeoffice.gov.uk/hmic/hmic.htm

Home Office Police Reform
http://www.policereform.gov.uk

National Black Police Association
http://www.nationalbpa.com

National Crime Intelligence Service
http://www.ncis.co.uk

Police Service UK
http://www.police.uk

Prisons and Probation (NOMS)

Her Majesty's Chief Inspector of Prisons for England and Wales
http://www.homeoffice.gov.uk/justice/prisons/inspprisons/index.html

Her Majesty's Prison Service for England and Wales
http://www.hmprisonservice.gov.uk

Home Office – National Offender Management Service
http://www.homeoffice.gov.uk/inside/org/dob/direct/noms.html

Home Office – Prisons
http://www.homeoffice.gov.uk/justice/prisons/index.html

Home Office – Probation
http://www.probation.homeoffice.gov.uk

Howard League for Penal Reform
http://www.howardleague.org

Howard League for Penal Reform in Scotland
http://www.howardleaguescotland.org.uk

Inside Out Trust
http://www.inside-out.org.uk

Northern Ireland Prison Service
http://www.niprisonservice.gov.uk

Northern Ireland Probation Board
http://www.pbni.org.uk

Prison News
http://www.fpe.org.uk

Prison Reform Trust
http://www.prisonreformtrust.org.uk

Prisons.org
http://www.prisons.org.uk

Prison and Probation Ombudsman
http://www.ppo.gov.uk

Scottish Prison Service
http://www.sps.gov.uk

World Health Organisation — Health in Prisons Project
http://www.hipp-europe.org

Restorative Justice

Mediation UK
http://www.mediationuk.org.uk

Home Office — Restorative Justice
http://www.homeoffice.gov.uk/justice/victims/restorative

Restorative Justice Ireland Network
http://www.extern.org/restorative

Restorative Justice Consortium
http://www.restorativejustice.org.uk

Restorative Justice Online
http://www.restorativejustice.org

Youth Justice Board — Restorative Justice
http://www.youth-justice-board.gov.uk/PractitionersPortal/
   PreventionAndInterventions/RestorativeJustice

Youth, Youth Crime and Youth Justice

Barnardos
http://www.barnardos.org.uk

Centre for Adolescent Rehabilitation (C-FAR)
http://www.c-far.org.uk

Children's Rights Alliance
http://www.crae.org.uk

Edinburgh Study of Youth Transitions and Crime
http://www.law.ed.ac.uk/cls/esytc

National Association for Youth Justice
http://www.nayj.org.uk

NSPCC
http://www.nspcc.org.uk

Save the Children
http://www.savethechildren.org.uk

Shaping the Debate
http://www.shapethedebate.org.uk

Youth Justice Board
http://www.youth-justice-board.gov.uk

Victimology

National Centre for Victims of Crime
http://www.ncvc.org

World Society of Victimology
http://www.fh-niederrhein.de/fb06/victimology/index.html

Victim Support
http://www.victimsupport.com

Appendix 7: Further Reading

Criminological Theory

Criminological Theory is probably the best place to start with your further reading as it will provide you with a basis to explore many of the key issues and debates within the discipline. The following texts are a good place to start:

Lilly, J.R., Cullen, F.T. and Ball, R. (2002) *Criminological Theory: Context and Consequences*, 3rd edn. London: Sage.

Tierney, J. (1996) *Criminology: Theory and Context*. London: Prentice Hall.

Vold, G.B., Bernard, T. and Snipes, J.B. (2002) *Theoretical Criminology*, 5th edn. Oxford: Oxford University Press.

The Criminal Justice System

We have discussed various aspects of the criminal justice system throughout this book and some of the texts identified here are referred to in previous chapters. They are a good basis on which to begin to develop your understanding of this key aspect of your degree. It is important that you remember that legislation relating to crime and disorder is constantly changing and you need to keep up to date with these changes. The following are examples of accessible texts but there are many more that deal with the topic:

Cavadino, M. and Dignan, J. (2002) *The Penal System: An Introduction*, 3rd edn. London: Sage.

Davies, M., Croall, H. and Tyre, C.J. (1998) *Criminal Justice: an Introduction to the Criminal Justice System of England and Wales*. London: Longman.

McConville, M. and Wilson, G. (2002) *The Handbook of the Criminal Justice Process*. Oxford: Oxford University Press.

Home Office Reports and White Papers that discuss reform of the criminal justice process are also of value. These are available on the Home Office website (http://www.homeoffice.gov.uk)

Policing

Police studies are increasingly a significant part of many Criminology degree programmes and there are a number of useful texts available that provide a good overview of the history and development of policing.

Johnston, L. (2000) *Policing Britain: Risk, Security and Governance*. Harlow: Longman.

Rawlings, P. (2002) *Policing: a Short History*. Cullompton: Willan.

Reiner, R. (2000) *The Politics of the Police*. Oxford: Oxford University Press.

Again, the Home Office produce research papers and reports relating to policing and these are an equally valuable source (http://www.homeoffice.gov.uk).

Criminal Justice and Punishment

There have been some important reports written in relation to reform of the criminal courts and the sentencing process. These are easily accessible and provide some useful information in relation to the way this aspect of criminal justice is changing. One report is cited below; this is the basis on which much of the change and restructuring of the court system in England and Wales is taking place.

Hough, M. and Jacobson, J. (2003) *The Decision to Imprison: Sentencing and the Prison Population*. London: Prison Reform Trust.

Hudson, B. (2003) *Understanding Justice*. Buckingham: Open University Press.

Sanders, A. and Young, R. (2000) *Criminal Justice*, 2nd edn. London: Butterworth.

A Review of the Criminal Courts of England and Wales by The Right Honorable Lord Justice Auld, September 2001 http://www.criminal-courts-review.org.uk

In relation to sentencing in the courts a second report can be accessed electronically and again provides background to the development of policy.

Making Punishment Work: A Review of the Sentencing Framework for England and Wales July 01, http://www.homeoffice.gov.uk/docs/halliday.html

'Race', Crime and Justice

There has been considerable discussion in relation to the subject of 'race' and the criminal justice process. Numerous reports have focused on the issue and there are some useful texts that will enable you to develop your understanding.

Bowling, B. and Phillips, C. (2002) *Race, Crime and Criminal Justice*. London: Longman.

Cook, D. and Hudson, B. (1993) *Racism and Criminology*. London: Sage.

Hudson, B. (ed.) (1996) *Race Crime and Justice*. Aldershot: Ashgate.

Crime Prevention

For some time the issue of crime prevention has become increasingly important. Government policy has focused considerable attention on the role communities can play in this as well as the way in which agencies can work together to impact on crime and disorder.

Crawford, A. (1998) *Crime Prevention and Community Safety: Politics, Policies and Practices*. London: Longman.

Hughes, G. (1998) *Understanding Crime Prevention: Social Control Risk and Late Modernity*. Buckingham: Open University Press.

Hughes, G., McLaughlin, E. and Muncie, J. (eds) (2002) *Crime Prevention and Community Safety: New Directions*. London: Sage.

General Introduction

In addition to the above texts and those identified earlier in the book relating to specific topic areas, there a number of good texts that provide a general introduction and background to the discipline of Criminology.

McLaughlin, E., Muncie, J. and Hughes, G. (eds) (2003) *Criminological Perspectives: a Reader*. London: Sage.

McLaughlin, E. and Muncie, J. (2001) *The Sage Dictionary of Criminology*. London: Sage.

Maguire, M. Morgan, R. and Reiner, R. (2002) *The Oxford Handbook of Criminology*. Oxford: Oxford University Press.

Glossary

This glossary contains a combination of academic terms and criminological terms, it is not an exclusive listing and more in-depth definitions and discussions of these terms can be found in other literature. Examples of this can be found throughout this book and include: *The Sage Dictionary of Criminology* (McLaughlin and Muncie, 2001), *Oxford Handbook of Criminology* (Riener et al., 2004) and texts on criminological theory.

Anomie	Ethical normlessness or deregulation, associated with the work of Durkheim and Merton.
Behaviour modification	Attempts to alter behaviour [of offenders] using learning theories.
Biological Criminology	The belief that certain people inherit a genetic or physiological predisposition to commit crime. Lombroso and Ferri are early proponents of this view.
Chicago School	School of thought linking environmental factors and crime.
Classicism	Theoretical perspectives that are underpinned by free will and rational choice. Seeks to develop a criminal justice system in which punishment is proportionate to the criminal act and acts as a deterrent. Key here is individual choice, the common interest and the social contract.
Community sentences	A range of penalties imposed for offending behaviour that does not involve custody. Often seen as alternatives to custody.
Conflict Theory	Often contrasted with the positivist view that there is a consensus in society. Involves conflict based on group identity, class identity or cultural identity. Often seen as relating to power and authority and attempts to understand the way in which the criminal law serves the interests of specific groups in society.
Corporate crime	Acts punishable by legislation that are the result of deliberate actions taken by formal organisations. Usually includes: large companies, businesses and corporations. Not to be confused with 'white collar crime'.
Crime control	The view that the main role of criminal justice agencies is to control crime. The primary function of the criminal justice

	system is the apprehension, conviction and punishment of offenders.
Criminal justice	The process through which the state deals with unacceptable behaviour. Involves arrest, trial, conviction and punishment of offenders. The agencies involved on this constitute 'the criminal justice system'.
Cultural Criminology	A recent perspective that emphasises the importance of style, image, representation and meaning in the construction of crime and crime control.
Cybercrime	Electronic communications used to commit illegal acts. Typically involves the Internet and web-based information and communication technologies.
Decarceration	A deliberate move away from imprisonment as the predominant penal sanction.
Delinquency	Loosely used to refer to youthful misbehaviour.
Deterrence	The use of sanctions to prevent criminal activity. Seeks to demonstrate the penalties for criminal acts outweigh the benefits to the offender.
Deviance	A term used to describe those acts that deviate from accepted social norms. Often contested, as the concept of deviance is relative rather than absolute.
Deviancy amplification	When media, public, criminal justice agencies and the state react to non-conformity in such a way that rather than controlling deviancy they increase it.
Discrimination	The unfavourable treatment of individuals or groups based on criteria such as age, gender, sex, race, culture, disability, ethnicity, language, social class, disability, sexual preference or any other inappropriate criteria.
Due process	The administration of justice according to legal rules that are transparent to the public and are seen as fair and just.
Ethnography	A research method based on the study of small groups, situations and contexts. Emphasises the importance of how individuals interpret and socially construct their world.
Fear of crime	Anxiety caused by the belief that individuals or groups are in danger of becoming victims of crime. This fear can be rational or irrational.
Feminist Criminologies	Analysis using feminist or critical theories to question the place of gender/sex in understanding crime and justice.

Focus groups	Interviews involving a number of individuals who discuss a specific topic supported by a facilitator. The data collected through the interaction that takes place are used to inform and develop research projects.
Folk devil	Seen as the stereotypical 'trouble maker' or 'drain on societies resources' and used to demonstrate the failings in society. Presented in a stylised manner by the mass media.
Functionalism	A perspective that is based on the understanding that society is structured in a specific way and sees crime and deviance as 'social facts'. These 'facts' are seen to perform a function in maintaining the smooth running of society. Functionalists would reject the view that criminals are pathological or abnormal.
Genetics	An attempt to identify the biological source/s of criminal and anti-social behaviour.
Hate crime	Crimes committed against individuals, groups or property and motivated by hatred, bias or prejudice. This is usually based on race, ethnicity, gender, sexual orientation or religion.
Hedonism	The pursuit of pleasure at the expense of rationality. Can be associated with risk and excitement in relation to crime.
Hidden crime	Crimes that fail to appear in official statistics, go unreported or are under-reported.
Incapacitation	One justification for punishment that seeks to prevent re-offending by the physical or geographical removal of an individual from the opportunity to commit crime. This can include imprisonment, banishment, or amputation of limbs with the death penalty being the ultimate form of incapacitation.
Incarceration	Confining deviant populations into specialist institutions for punishment or treatment.
Informal justice	This is an attempt to deal with criminal and anti-social behaviour without recourse to the formal processes of the criminal justice system.
Labelling	The sociological understanding of the way in which certain groups and individuals are classified and categorised by others in society. The stereotyping leads to the labelled individuals to respond to their label which reinforces the

	self-perception of individuals and the perception of society to those groups.
Longitudinal studies	A study collected over time using the same research participants, sometimes at key points in their life course. Sometimes referred to as cohort studies.
Moral panic	Reaction to behaviours, people or groups that is disproportionate to any real threat posed to society's values and is based on stereotypical, media representations. Often leads to calls for greater social control.
Official crime statistics	Based on data recorded by the police and the Courts. Indicates the extent of crime. There are debates that question the relevance of these figures in relation to the amount of actual crime committed.
Participant observation	Collection of data through active participation with groups or individuals who are the subject of the study.
Positivism	This theoretical approach emerged during the nineteenth century. It argues that social relations and events can be studied using methods that are drawn from the natural sciences. Positivists seek to explain and predict future behaviour.
Probation	Supervision of offenders in the community by officers of the Court.
Reflexivity	Monitoring and reflecting on all aspects of research, from initial idea to writing final report.
Sub-culture	Often applied in relation to delinquency. Refers to different values that challenge the mainstream norms of the dominant culture.
Underclass	Refers to those who are seen to be poor but have adopted values that lack morality. Are 'feckless' or 'undeserving'.

Bibliography

Alvesalo, A. (2003a) *The Dynamics of Economic Crime Control*, Vol. 14. Espoo, Finland: Poliisiammattikorkeakoulun tutkimuksia.

Alvesalo, A. (2003b) 'Economic crime investigators at work', *Policing and Society*, Vol. 13 (2), 115–38.

Bandura, A. (1977) *Social Learning Theory*. London: Prentice Hall.

Bandura, A. (1986) *Social Foundations of Thought and Action: A Social Cognitive Theory*. London: Prentice Hall.

Becker, H. (1963) *Outsiders: Studies in the Sociology of Deviance*. New York: Free Press.

Cavadino, P. and Dignan, J. (2002) *The Penal System*, 3rd edn. London: Sage.

Chibnall, S. (1977) *Law and Order News: An Analysis of Crime Reporting in the British Press*. London: Tavistock.

Cohen, S. (1973) 'The failures of criminology', *The Listener*, 8 November.

Cohen, S. (1988) *Against Criminology*. Oxford, New Brunswick, NJ: Transaction Books.

Croall, H. (1998) *Crime and Society in Britain*. Harlow: Longman.

Currie, E. (1985) *Confronting Crime: An American Challenge*. New York: Pantheon.

Dalton, A. (2000) *Consensus Kills: Health and Safety Tripartism – a Hazard to Workers' Health?* London: AJP Dalton.

Dawson, S., Willman, P., Bamford, M. and Clinton, A. (1988) *Safety at Work: The Limits of Self-Regulation*. Cambridge: Cambridge University Press.

Dennis, N. and Erdos, G. (1992) *Families Without Fatherhood*. London: Institute of Economic Affairs.

Ditton, J. (1979) *Contrology: Beyond the New Criminology*. London: Macmillan.

Downes, D. and Rock, P. (1998) *Understanding Deviance: A Guide to the Sociology of Crime and Rule Breaking*, 3rd edn. Oxford: Oxford University Press.

Downes, D. and Rock, P. (2003) *Understanding Deviance*, 4th edn. Oxford: Oxford University Press.

Earle, R., Newburn, T. and Crawford, A. (2003) 'Referral Orders: some reflections on policy transfer and "what works"', *Youth Justice*, Vol. 2, No. 3, 141–50.

Fairbairn, G. and Winch, C. (1996) *Reading, Writing and Reasoning: A Guide for Students*, 2nd edn. Milton Keynes: Open University Press.

Farrington, D. (2002) 'Developmental and risk-focused prevention', in M. Maguire, R. Morgan and R. Reiner, *The Oxford Handbook of Criminology*, 3rd edn. Oxford: Oxford University Press.

Farrington, D.P. (1992a) 'Juvenile delinquency', in J.C. Coleman (ed.), *The School Years*, 2nd edn. London: Routledge & Kegan Paul.

Farrington, D.P. (1992b) *British Journal of Criminology*, Vol. 32, No. 4 (Anthum).

Foucault, M. (1991) *Discipline and Punish: The Birth of The Prison*. London: Penguin.

Garland, D. (1985) *Punishment and Welfare: A History of Penal Systems*. Aldershot: Ashgate.

Garland, D. (1990) *Punishment and Modern Society: A Study in Social Theory*. Oxford: Clarendon.

Garland, D. (2001) *The Culture of Control: Crime and Social Order in Contemporary Society*. Oxford: Oxford University Press.

Giddens, A. (1997) *Sociology*, 3rd edn. Cambridge: Polity Press.

Goldson, B. and Jamieson, J. (2002) 'Youth crime, the "parenting deficit" and state intervention: a contextual critique', *Youth Justice*, Vol. 2, No. 2, 82–99.

Hall. S., Critcher, C., Jefferson, T., Clarke, J. and Roberts, B. (1978) *Policing the Crisis: Mugging the State and Law and Order*. Basingstoke: Palgrave MacMillan.

Herrnstein, R.J. and Murray, C. (1996) *The Bell Curve: Intelligence and Class Structure in America*. London: Free Press.

Hester, S. and Eglin, P. (1992) *A Sociology of Crime*. London: Routledge.

Home Office (1997) *No More Excuses*. London: Home Office.

http://www.homeoffice.gov.uk/rds/pdfs04/dpr26.pdf

Hudson, B. (2003) *Justice in the Risk Society*. London: Sage.

Lilly, J.R., Cullen, F.T. and Ball, R.A. (2002) *Criminological Theory: Context and Consequences*, 3rd edn. London: Sage.

MacDonald, R. (ed.) (1997) *Youth, the 'Underclass' and Social Exclusion*. London: Routledge.

Maguire, M., Morgan, R. and Reiner, R. (2002) *The Oxford Handbook of Criminology*, 3rd edn. Oxford: Oxford University Press.

McLaughlin, E. and Muncie, J. (eds) (1996) *Controlling Crime*. London: Sage.

McLaughlin, E. and Muncie, J. (eds) (2001a) *Controlling Crime*, 2nd edn. London: Oxford University Press.

McLaughlin, E. and Muncie, J. (2001b) *The Sage Dictionary of Criminology*. London: Sage.

Mead G.H. (1934) *Crime Law and Social Science*. New York: Harcourt, Brace Jovanovich.

Merton, R.K. (1938) 'Social structure and anomie', *American Sociological Review*, 3, 672–82.

Mills, C. Wright (1970) *The Sociological Imagination*. Oxford: Oxford University Press.

Morris, L. (1995) *Social Divisions, Economic Decline and Social Structural Change*. London: UCL Press.

Muncie, J. (1999) *Youth and Crime: A Critical Introduction*. London: Sage.

Muncie, J. (2002) 'A new deal for youth? Early intervention and correctionalism', in G. Hughes, E. McLaughlin and J. Muncie (eds), *Crime Prevention and Community Safety New Directions*. London: Sage.

Muncie, J. (2004) *Youth and Crime*, 2nd edn. London: Sage.

Muncie, J. and McLaughlin, E. (eds) (2001) *The Problem of Crime*, 2nd edn. London: Sage.

Muncie, J. and Wilson, D. (2004) *Student Handbok of Criminal Justice and Criminology*. Cavendish: London.

Murray, C. (1984) *Losing Ground: American Social Policy 1950–1980*. New York: Basic Books.

Murray, C. (1990) *The Emerging British Underclass*. London: Institute of Economic Affairs.

Murray, C. (1994) *Underclass: The Crisis Deepens*. London: Institute of Economic Affairs.

Newburn, T. (2003) *Crime and Criminal Justice Policy*, 2nd edn. Harlow: Pearson education Limited.

Observer (2004) 'Britain takes a crash course in happy families', http://observer.guardian.co.uk/politics/story/0,1153456,00.html

Park, Robert and Burgess, E.W. (eds) (1925) *The City*. Chicago: University of Chicago Press.

Parker, H. (1974) *View From the Boys: A Sociology of Down-town Adolescents*. Newton Abbot: David Charles.

Pearson, G. (1983) *Hooligan*. London: MacMillan.

Phillipson, M. (1971) *Sociological Aspects of Crime and Delinquency*. London: Routledge and Kegan Paul.

Pitts, J. (2001) 'Korrectional Karaoke: New Labour and the zombification of youth justice', *Youth Justice*, Vol. 1, No. 2, 3–16.

Reith, C. (1943) *A Short History of the Police*. Oxford: Oxford University Press.

Rudestam, K.E. and Newton, R.R. (2001) *Surviving Your Dissertation*. London: Sage.

Sharrock, W. (1984) 'The social realities of deviance', in R.J. Anderson, and W. Sharrock (eds), *Applied Sociological Perspectives*. London: Allen and Unwin.

Smith, R. (2003) *Youth Justice, Ideas, Policy, Practice*. Cullompton: Willan Publishing.

Smith, D. and Tombs, S. (1995) 'Beyond self-regulation: towards a critique of self-regulation as a control strategy for hazardous activities', *Journal of Management Studies*, Vol. 35, No. 5, 619–36.

Storch, R. (1975) '"The Plague of Blue Locusts" Police Reform and Popular Resistance in Northern England 1840–1857', XX *International Review of Social History*, 61.

Tierney, J. (1996) *Ccriminology: Theory and Context*. London: Prentice Hall.

Vold, G.B., Bernard, T. and Snipes, J.P. (2002) *Theoretical Criminology*, Oxford: Oxford University Press

Walklate, S. (1989) *Victimology: The Victim and the Criminal Justice Process*. London: Unwin Hyman.

Webster, C., Simpson, D., MacDonald, R., Abbas A., Cieslik, M., Shildrick, T. and Simpson, M. (2004) *Poor Transitions, Social Exclusion and Young Adults*. Bristol: Policy Press, Joseph Rowntree Foundation.

Williams, K. (2004) *Textbook on Criminology*, 5th edn. Oxford: Oxford University Press.

Winlow, S. (2001) *Badfellas: Crime, Tradition and New Masculinities*. Oxford: Berg.

Young, J. (1999) *The Exclusive Society*. London: Sage.

Index